#MYEPICYEAR

RISING FROM THE ASHES

BRITTNEY OLIVER CHC

Copyright © 2019
Brittney Oliver CHC
#MYEPICYEAR
Rising From The Ashes
All rights reserved.

No part of this publication may be reproduced, distributed, or transmitted in any form or by any means, including photocopying, recording, or other electronic or mechanical methods, without the prior written permission of the publisher, except in the case of brief quotations embodied in critical reviews and certain other non-commercial uses permitted by copyright law.

Brittney Oliver CHC

Printed in the United States of America
First Printing 2019
First Edition 2019

10 9 8 7 6 5 4 3 2 1

This book is dedicated to every single person who chooses to transcend circumstance. You gave my fight purpose and that was worth every minute.

~ Coach Britt

Table of Contents

Introduction 1
Chapter 1 5
  Yup… I'm A Life Coach
Chapter 2 19
  My Hand Of God Moment. Rock Bottom
Chapter 3 31
  Understanding You
Chapter 4 41
  Awaken
Chapter 5 53
  Awaken…. Again
Chapter 6 75
  Your True Value.
Chapter 7 93
  Running At Fear.
Chapter 8 107
  Purpose From Pain
Chapter 9 119
  A Warrior's Blueprint
Chapter 10 135
  Finding The Light In Darkness
Chapter 11 147
  Fire Walk
Resource List 165

Introduction

"Reality is merely an illusion. Albeit a very persistent one."

~Albert Einstein

In a space of a second, your reality can change for better or worse. Most people envision a deathly car accident, or a paramount decision of sobriety or change. What if that moment was more of an awakening? A true realization of who you truly are and what you are truly capable of? Just as a person who's always been colorblind knows no color, we too have been molded throughout our lives to see in black and white. The real challenge of living is navigating through our predisposed beliefs and spotty vision, so that we may find a color pallet which once only existed in our dreams; that is seen only in faith. Faith doesn't have to be religious per se. Faith is the unwavering belief in something that we have not yet seen. Believing in something which you cannot see takes a certain degree of confidence and that confidence is capable of riding the hardest waves in life.

Every person is capable of greatness but first, we must properly define greatness. Great leaders throughout history were obviously great because of their extraordinary contributions, influence and large reach throughout the world as a whole. Some were most definitely designed to become "obviously great," but what about everyone else? What about all the people who won't go on to lead the masses, die for selfless victories or sacrifice for the betterment of mankind? How then can everyone truly be designed for greatness? The answer is simple, and it lies in purpose.

Greatness in the grandiose sense is subjective. While not to downplay the incredible impact that so many great leaders have made; the idea that to be great, we must be big is an articulation of man's perspective. Each individual has a purpose in this world and therefore, was designed to become great. Some may only touch the lives of few, but who are we to gauge? What are those few meant to do in this life, and what did that person who reached them actually contribute? If Gandhi had been empowered by one person and this one person knew only 5 people in their lifetime, was that person not a contributor to the greatness that Gandhi became and all that manifested from him? In that aspect, was Gandhi's mentor less great than he? Or was he an equal piece of the puzzle, living out his purpose, following his destiny and becoming his own greatness? You see, we cannot predict the impact that we will have on our world and because of that, we are gifted the extreme satisfaction of knowing that as long as we seek to heal and become the best soul versions of ourselves, we will become so great in purpose. By doing this consciously and self-aware, we will also benefit from a life well lived. A story in which we write the ending.

This book wasn't written as an optimistic mirage of all that can become of life, nor was it written as a melancholy, "self-help" book that is designed to focus on your past and personal imperfections. This book was crafted with the intention of being a manual so to speak. An autobiographical, soul written documentation of personal understanding and purpose throughout the most intense year of my life. Documenting the entire process of suffering, fighting and finding purpose from pain from beginning to end. We need to ride the whole wave with a high level of consciousness in order to create the ending. But first, we must start at the very beginning.

BRITTNEY OLIVER CHC

Chapter 1
Yup... I'm A Life Coach

I'm Coach Britt; a certified health and life coach and founder of a successful, growing elite coach training school. That's not exactly how I should be starting out a book that fully documents some of the hardest and most shameful moments of my life but hey; transparency is the key to transformation I always say! I am a leader, a mother, an author and generally someone who puts herself out into the world. We are all human beings working through this life, and while I am blessed to do what I do, I am not beyond mistakes nor am I untouchable. Sometimes, life just happens at very inconvenient times and all you can do is simply to embrace the process. So, as a coach, a leader, a mother and a human that is exactly what I did.

I got into this world of holistic wellness and personal development through a lifetime of struggling with chronic anxiety and depression. It is true that suffering breeds evolution. So, our personal stories are quite unique and individualized. I believe that this is because we as individuals have a unique calling in this world and therefore, the topography of the journey is different for each of us. Some need a steep path while others have jagged peaks or low, and muddy valleys. The design of your journey is specifically to mold you for your individual calling. My journey started with fear. My early years were filled with fear and a lot of darkness. A darkness reminiscent of a never-ending bad dream.

My earliest memories of constant fear started at about 4 years old. Most of the time, it probably looked like regular childlike behavior; fear of being anywhere else besides my home or anxiety about being watched by anyone besides my parents (which didn't happen often). By first grade, it became so obvious that something more complex was really going on, as I can vividly remember crying almost every day at school. Not in front of my teachers but alone, enclosed at a corner in an attempt to conceal it. I would experience torturous guilt and worry in a very irrational manner. One specific memory that stands out to me was a time which my dad had walked me to school (we didn't have a car until I was 12), and in the course of it all we forgot my lunch. He dropped me off and ran to the store to pick up something for me to eat. After going out to meet him and get my lunch, I was overcome with guilt. I bawled my eyes out the rest of the school day due to an overwhelming feeling that made no rational sense. I felt so guilty, and so sad. I felt that he had done something too nice for me. I worried about my parent's money and felt sorry that my father had to go to the store all for me. My mother worked very hard and they provided for all of my basic necessities, so my emotional response was confusing at best. My parents were oblivious to these occurrences and I sure didn't make a point to tell them anything.

Eventually, my teacher would have a meeting with them while expressing her concern for my constant "secretive" crying, and my parents being good parents would try to address it with me. They did what any concerned parent would do; they asked me WHY I was upset, in an attempt to figure out WHAT was going on, causing me so much pain. They were as dumbfounded as I was, and gave their best effort at figuring out a 7yo. My response; I blamed it on a girl at school who would sometimes pull my hair.

Why did I blame her? Because my parents and teacher needed an answer so desperately. An answer from a little girl who was experiencing a hell that most grown adults have trouble articulating. A hell that overwhelmed almost every moment of my childhood life. They did their best and told my teacher about the girl who pulled my hair in an attempt to eradicate the issue, and my teacher then discussed this with the girl and that was that. I continued to struggle in secret, especially because I had no more answers or explanations. Now that everyone had "fixed" the problem for me, there was surely no way that I could let them know that it still existed! I couldn't possibly ask them to worry about me knowing that I didn't have an answer as to how I could be helped, and so I continued to tuck it away as I suffered in silence, alone.

In 3rd grade, I tested at a college level for my creative writing abilities and articulation. I had a gift for expressing emotions and pictures through my words. Because of this, they wanted to send me to a special school 1 day a week. The advanced school wanted to first look at my 3rd grade test scores for one of the required tests at that time, in order to approve of my attendance. This scenario peaked my anxiety tenfold! I was so terrified about leaving on a bus to go somewhere unknown to me that I purposely failed my test in order to assure my rejection from the program. My anxiety combined with my fear of the unknown made what would have been a great opportunity non-negotiable. There was no way that I was getting on that bus.

This was a rough year. I went through numerous strange and exhausting phases such as the time where I couldn't swallow food. It was a stress response that caused me to constantly choke. It was terrifying and played into my fear of dying. As an adult, I can't quite wrap my brain around it rationally, but as a child, I can clearly tell you that it was very real. Instead of telling anyone, I would remedy the situation by only eating soft foods. Later on, I was able to look back on this and see the correlation with my hyper-anxiety, however, at the time I just felt like a weirdo. I must have gone weeks avoiding solid food. Every time I would have to eat, my heart would pound and I would become overwhelmed with trepidation. I was way too embarrassed and confused to ever share this struggle with anyone, so I endured until it eventually subsided.

By 4th grade, I started experiencing bouts of long-term anxiety and depression which manifested into a constant fear of death. It got so bad that I couldn't even watch certain types of shows on TV. All medical shows or anything dealing with death or dying would throw me into an all-night sleepless frenzy. I was paranoid that those things were going to happen to me. I couldn't shake out of my constant dread. All I could do was pray. Almost every night, I would find myself in a hopeless panic, so I would pray. Only God could possibly know what I was going through. I'd pray for safety, and I'd also pray not to die. When I would finally start to come down from my panic attacks, I'd cry as the tears always signaled the release. They marked the come down from a state of sheer terror, and I would cry and shake uncontrollably until I eventually fall asleep. It escalated so badly that there were multiple occasions where I wouldn't go to school for days. I'd tell my parents that my "tummy hurt" and that "I didn't feel good." It wasn't really a lie! Severe panic, depression and stress comes with a multitude of physical manifestations, especially with your tummy, but I needed an excuse. An explanation for something that felt impossible to me, so I told them that I was sick.

These instances happened off and on for a couple of years until it came to a massive head during my junior year in high school, and I suffered a full-blown nervous breakdown. I was going through some teenage relationship drama which likely acted as a switch. I started experiencing constant, non-stop waves of intense panic followed by the darkest bouts of depression I'd ever had. The anxiety felt like a fear that words can't even describe, and the depression I can only explain as a sick world of death. Like being lost in a cave of skeletons and demons with no light in sight, I didn't leave my bed for a week. I almost lost my job and couldn't go to school. My friends started calling to check on me and again, I would tell people that I was sick. What was I supposed to say, "I'm terrified to leave my house?" or "I'm so depressed that I can't fathom human interaction?" When I finally tried to get back to school, it didn't go well. One specific day, I was in math class and my panic attacks kicked in so badly that my vision blurred and I lost most feeling in my arms. I asked to be excused, and while walking through the hallway I was approached by a concerned friend. At this point I lost it, so I cried, hyperventilated and explained that I couldn't feel my arms or face. They took me to the office and had another friend drive me home.

How the hell could I live like this? The depression piece put me in such a dark black world of my own that even those close to me couldn't understand and because of this, I couldn't really let them in. To this day, my heart hurts so deeply for those who contemplate or attempt to take their lives as well as for those who fight one day at a time. Because I know what that darkness looks like, and I have experienced what immense fear feels like, I have so much compassion for those who are still living in the darkness. It is a place that only exists in nightmares except the nightmare is your mind and it goes wherever you go. I don't know how I began living a regular routine again. It was a slow and gradual process and I refused to keep hiding away for fear that I'd never be able to crawl out. While I eventually became functional again, I knew that I wasn't healed.

I spent years riding this rollercoaster. In my late teens, early 20's, I would sometimes self-medicate with alcohol. I had a very hard time in crowded public places despite being a highly extroverted person, and I found out that I could drink to alleviate some of that angst. While I have never been a frequent drinker, I later got to a point where I had to realize that my ability to self-regulate the amount that I would drink was not so good. Regardless of the fact that alcohol would only exacerbate my anxiety and depression, there were times where I felt that it was my only way to participate in life. I had missed so many moments. I missed both my Mom's graduation from college and my little sister's graduation from high school due to my inability to be in the crowded facility. Through my early to mid 20's, I experienced a lot of agoraphobia which is a fear of public places or leaving your house. I would see myself struggle so badly with things as simple as grocery shopping. All that I could imagine was passing out in the middle of the store. Before making a trip to grocery shop I would make my list, map out my quickest exit and fly like a bat out of hell. I'd pray that it wasn't too crowded and that I wouldn't get stuck behind the person with 100 coupons! Life was a lot of work and I was burnt out. Interestingly though it was all I'd really ever known. I'd lived this way since I was 4 years old and as sad as that whole scenario seems, it is probably what saved my life or at least became the catalyst for my purpose.

During my nervous breakdown experience in my junior year, I had begged my dad to put me on anti-depressants. I was desperate and at that point, it was all that I was ever taught to believe could help me. For whatever reason, this never ended up happening. I am now incredibly grateful that I was never medicated. I don't believe that I would have discovered how to truly heal, control or understand this piece of my life had I tried to mask it. I may never have dug deeper; never discovered the true roots. I persevered because I didn't know another way and that perseverance eventually led me to seek positive answers. Those answers then inspired an unrelenting fight to concur the very thing that I had spent my whole life feeling helpless to.

My healing journey did not happen overnight. I still struggle with occasional bad days but overall, I don't even know that level of depression or anxiety anymore. It's funny what happens to you when you finally stop suffering. You start living. I thrive off empowering individuals to self-heal, so that they too can truly live. I had spent the majority of my life surviving and when I finally started experiencing life without constant fear and pain, it was like advancing into a whole new world; a universe of limitless potential and light amongst the darkness. I totally refused to take this new life for granted. I greatly desired to use my own journey in life to help others. I can assure you that this road has been a rocky one but none the less, one worth sharing. I thrived for so many years before being challenged again, and this time to a new degree. Life tends to ebb and flow, and I have found out that there will be periods throughout your entire life's journey which levels you and tests your ability to fight. My fight wasn't over after high school, in some ways it was just beginning. Most people will associate me with a bubbly, positive personality or intense passionate energy, but those pieces of me were molded from pain and motivated by an extreme desperation to heal and live. I fought for the person which I have become and I won't ever stop fighting. I realized earlier on that if I wanted to live fully, and relieve the weight of pain from my body that I would have to get stronger and start lifting some boulders and by doing that, maybe others would follow. By sharing my pain, fight and successes, maybe other people would believe that they too can fight and eventually succeed.

You may not yet have been to a place where you can feel inspired by the healing stories of others. You may be in that mode of desperation or survival but ultimately, I desire for you to find hope in true healing. We are not our story; we are who we choose to become because of it. Look at any child and any baby, we all came into this world as innocent sponges, ready to absorb the environment around us with absolutely no control over it. Every person is a result of their experiences as well as their choices. Everyone has a story. I would be willing to bet that no bomber, school shooter, murderer, bully, abuser or criminal has an individual story that doesn't include a lot of suffering. By no means am I saying that the horrendous actions that take place daily are excused. Unresolved pain and suffering only lead to more pain and suffering. It is a highly venomous energy. On the other hand, so is love and compassion. Sometimes it only takes one person, in one moment to completely change the trajectory of a human life. This is why it is so crucial that we evolve from where we are and pursue purpose in this world! Heal yourself and then help those around you do the same.

My passion for leading others out of suffering was what brought me to the world of holistic health and life coaching, and while I have spent days and sometimes weeks throughout this journey feeling like a failure or a hypocrite, I believe that true transparency is required in order to achieve true transformation. To me, this was a non-negotiable commitment to myself, God and those who follow me. While I may fall down at times, I can promise you that you will always see me stand back up. There is nothing in this world worthy of taking away your life, or essentially your purpose. Although it took me a long time until I was able to fully live, the life that I have fought for was worth every moment.

Don't just "not give up," nurture yourself and those around you until you find that fire that eventually leads you to concur the mountain. I promise that it is an incredible view.

Chapter 2
My Hand Of God Moment. Rock Bottom

There I was, handcuffed in the back of a cop car, shamefully listing my profession as a health and life coach to the inquiring officer. This night started like many others had; a busy day alone with the kids followed by a repetitive barrier in my relationship which led to a typical fight between my husband and I. Only this night would unbeknownst to me become the catalyst for the most painful and transformative year of my life thus far. On this warm May night, as I stood on what was currently the top of my game with business, leading so many people and blessed with 3 healthy kids, my personal life finally took a toll on me.

My husband and I had been battling some unrelenting issues for many years and it had all finally come to a head. I decided that I needed to heal with or without him and his participation. As a mom of 3, I started getting my ducks in a row. I had gone from praying that he'd finally get help to coldly caring less. The things that had affected us were poisonous. In as much as I was persevering, convincing myself that I was strong, I could do anything, I was okay, it wasn't true. Sometimes, being truly strong means being able to admit weakness, seeking help or letting others in. I carried way too much for way too long. The thing is, pain begets pain. When we don't healthily deal with the wounds which we carry, they become a ticking time bomb just waiting to spew shrapnel onto the world around us. That's exactly what happened with me.

I decided to have a drink, which turned into two. Soon after this my husband and I got into it big time. It was so bad. Nothing was getting better and I knew that it never would. I left and went on a drive by myself, parked in an isolated area and broke down to a friend over the phone, crying for 2 solid hours. I had essentially run away from home because I was finally losing it. I was already on the brink of leaving, I had already let things go too far but yet, here I was, questioning my own sanity and choices. Maybe it was really all me? Maybe this was all my fault? Was he right? Did I just "think wrong" and "see things wrong?" Normally I don't drink often and I sure don't drink in a car! I really don't drink and drive a car! But this wasn't a normal night for me in any way. I still had more alcohol unopened in the back seat with the groceries, and while on the phone with my friend I preceded to drink more.

I planned on sleeping in my vehicle in the dark parking lot which seemed more like a Hilton compared to going home that night. After an hour and a half feeling nothing but my emotions at the time, I got off the phone, sat there and prayed to God. I cried and prayed, telling God that I couldn't make a decision and that I couldn't be sure if staying or leaving would hurt my children more. I cried and begged him to make the decision for me as I was too weak and too unsure of myself. My marriage had been very mentally abusive, and it poisoned my mind and altered my reality. I could no longer trust what I saw or felt, so how could I make a decision as important as this? I remember years of being certain that I would never be anywhere else or with anyone else and yet, here I was. I wanted to run away into the mountains, just drive and let things go. I couldn't leave my kids, but I just wanted peace. My soul hurt and I couldn't take this feeling anymore! I felt almost dead for the first time in years. I had sworn that he loved me, so how did I feel like nothing to him? He made it very clear that my pain wasn't his problem and that keeping our family together wasn't worth the effort of getting help. I was stuck in total grief and so much confusion. What was I going to do and how could I possibly trust my decisions when I couldn't even trust my own sanity? I eventually got it together enough to decide that I'd better not sleep in the parking lot of the beach as it is patrolled often, so instead, I attempted to drive home. After deciding that I didn't want to go home, I parked in a more public place and started calling friends or family. At this point, I planned on having one of them come get me. As I was sitting in my driver's seat, parked and hysterically crying on the phone once again, the cops showed up. They had reason to believe that I was drinking and ran the usual protocol. Needless to say, I failed. I was cuffed (rather nicely I will add), read my rights and taken down to the jail. I got a cute yellow outfit to wear, and sat in the holding unit until my husband and his kind-hearted step dad came and posted my bail.

Although I had never had even so much as a speeding ticket in my life, what I had just done was bad, really bad. I hadn't dealt with my pain and now it was shooting beyond me. Thank God I got busted that night. Thank God I didn't hurt someone else. While the financial and driving punishments that accompany a

DUI are intense, they were merely inconvenient compared to the guilt and shame of a choice that led to an action that I could not take back. As a leader, a teacher and a lover of people, how could I be so selfish? The only reason I had been finger printed before this night was for the volunteer work that I had spent the previous year doing! Although this one night did not exemplify my typical character, and although my record was spotless, those things didn't absolve me from the consequences of my selfish moment. I didn't deserve nor desire to be relieved of my guilt over what I had done. That guilt for what could have happened, and all of the people I let down (weather they knew or not) was almost too much. I spent countless days dreading getting out of my bed. I would spend every night tossing and turning with an unforgiving pit in my chest. I cried more in that first two weeks than I had in ten years. I felt the need to punish myself in an attempt to alleviate at least enough guilt to allow me to function properly.

I had no idea how I would move through this, but one thing that I did know for sure was that the only thing worse than the mistake that I had made, would be to leave my life at that point. Not getting out of bed or giving up on life would be giving up on my children, students, friends and family. Each morning I'd say to myself "I don't want to get up." But I made sure not to say "I can't get up." That would have guaranteed that rock bottom was my new home instead of my new foundation; A foundation that I needed to rebuild. In my position, I had a long way to fall and it hurt like a mother fucker.

When this original incident had taken place, I wondered what God was communicating to me? Did he want me to stay? I would surely be losing my license for a while as well as facing hefty expenses. This did not make moving out and pursuing life as a single mom very feasible. Was I supposed to stick it out? Would things get better? It only took about 48hours for the answer to reveal itself. I knew from the bottom of my being that I needed to go. Through my pain I had allowed myself to fall to a level that could have taken away everything that meant anything to me. My children, my life's work and even my freedom. It also could have taken from someone else, completely innocent in all of this and I thanked God that it didn't. It became clear to me however that my prayers had been answered. It was surely time to go. Where there is a will, there is a way and with all the odds against me I started planning my way out with no license and very little money.

Two days later, I sat in a court room listening to the judge read all the charges of the other offenders. I absorbed so much about these people in that two hours. Their stories were their pain incarnate. I couldn't help but feel out of place. But was I really? My pain had gotten me here just as theirs had. I wondered who these people were as kids. What happened to them? What could have been had someone helped them heal instead of medicate? I realized that this experience had been given to me, which didn't mean that I wasn't responsible for my choices, it just meant that what's done was done and I needed to absorb the lesson now. So, I did. While the process was quick as far as legal stuff goes, it still lasted over 2 months. Just when I'd start to gain my footing and get back to life, I would be faced with another piece of the legal process which would remind me of one of the worst nights of my life. Mind you that during this DUI process, I was painfully wrapping up the end of a 5year marriage and running my active coach training business in an attempt to support myself and my boys alone. I was taking a mental, emotional and physical beating and all I could do was have faith and proceed with the process.

When I was technically sentenced, it included 3 days' work on the Sherriff's labor force. Since the Academy and all that I had going on at the time was in full effect and had started doing well, I was literally putting on a skirt and heels for a business meeting one day and grungy labor clothes the next day. I'd hop on a bus with a handful of other criminals and put in my work time. While this experience was hard, it taught me something valuable. It taught me that there is growth and healing in humility, but only if we are open to receiving the lesson. I was so open. I'd never been more open in my life, and boy I did learn. I eventually got through my DUI process. I also got through my divorce. The pain from all of it didn't dissolve overnight. In fact, it transformed me. By the one year mark of this event, I would no longer be the person whom I was before, but then again, I was never meant to be.

Where we go after a hard time or painful event is entirely up to us. Your perspective will determine your direction and your direction will determine your resiliency. I share all of this now not because it's easy, but because it's necessary. We are not our mistakes, but rather who we choose to become because of them. My mom called me for days saying, "Honey, you made a mistake but it does not define you." Good thing that she repeated this to me so many times because I couldn't believe it. Through counseling, I later realize that I possessed very low self-compassion and that amplified the shame of my situation. I struggled in believing that I was a worthy human after what I had done, which became yet another hurdle added to my run, but I was getting good at my jump.

I'd be lying if I said that sharing this piece of my story isn't hard. I have chosen to risk my professional position and reputation as well as taint the way people look at me indefinitely. I could have easily left this chapter out and still allowed the entire book to come together, but I knew that my journey served a purpose. I had to share my own lows, shame and mistakes. You aren't the only one who's screwed up or has made mistakes! I waited until age 33 to make the worst mistake of my life! I waited until I had built a life of leadership and expectation before I decided to throw myself under the bus!

My moment was not just a learning experience, it was a message. I was always the "strong" one. I self-soothed and rarely asked anyone for anything. I dealt with my own hurts and pains, and didn't even know how to really be vulnerable or "not okay." I call that night my "Hand of God moment." I had been dealing with my intense and long running pain alone, while thinking that I was taking care of it, but I wasn't. That incident revealed my unhealthy and painful state to everyone, such as my parents, friends, in-laws and of course law enforcement who got my full marital story on their handy dandy chest cams! I had to tell even more people since I had to explain my suspended license. The interesting thing about all of this was that while it added the weight of shame and guilt, it alleviated something. That something was the boulder of hurt and pain that I'd carried for miles, all alone. It wasn't so heavy anymore. I no longer carried it by myself. My close friends and family rallied around me, checking up on me and supporting me through the excruciating process of divorce with kids. I appreciated the love and acceptance given to me by those around me as I did not yet, poses it for myself. If I could do it all over again, I wouldn't have waited until my pain caused me to implode. I would have trusted more people with my situation and gotten into that counselor that I hadn't yet made time for.

With the past now unchangeable, I then trusted the healing process. I accepted my consequences as well as my blessings, so I committed to personal healing and a new level of growth. And most of all, I promised to use my story, no matter how hard it may be.

Chapter 3
Understanding You

Inner strength refers to one's ability to be resilient and vigilant in the pursuit of survival no matter the circumstances. The ability to bounce back, or rather bounce forward is a very important factor in life. Resilience has been touted as the single most beneficial trait when overcoming past traumas and damaging life circumstances. While I agree that this is true, I learned something else in my early 30's. You see, resilience had always been my forte because I was nothing if not for the fact that I was resilient. I remember getting unfairly fired from my main job at age 18 right after I'd signed a 6month lease on my first apartment. While I hung my head and wiped my tears that day, I woke up first thing the next morning and pursued finding a new job. I did this with vigilance until the mission was accomplished, and I rode an almost high the entire way. I always seemed to gain strength and momentum from adversity and because of this I thought that I had it completely made. I could get through anything in life and come out better than when I started. One particularly hard time, however, showed me that there was a second component to rising from the ashes and that was self-awareness.

Self-awareness is defined as a "conscious knowledge of one's own character, feelings, motives and desires." It exemplifies an individual's ability to evaluate their intentions and consciously navigate their feelings and behaviors, developing a better understanding of why they think, feel and act as they do and embracing the power to change and evolve. Self-awareness is now in my opinion, the most important personal trait. During what I call "My Epic Year," I tapped into a whole new level of self-awareness. I had been at my rock bottom and realized that this crumbling had happened after years of resilient fight. Constantly picking myself up from every knock down and hurdling every obstacle, and yet here I was face down in the gutter having fallen so far off of the mountain I had scaled. Life is a culmination of success and failure. I know that sometimes we will fall and sometimes the distance may be far, but why did I feel like I had put in so much fight, only to end up lower than I had started? After much thinking and praying, I concluded that this happened because I ran with resiliency, but failed to grow enough in the depths of my self-awareness. This is what I refer to as "Shooting in the dark." We fire hard but don't exactly understand our target. We pick ourselves up and sprint, but we don't have a clear idea what we are running at or why we believe this to be the actual way.

Much of our beliefs and behaviors are cultivated deep within the subconscious mind which is 30,000 times more powerful than the conscious mind! Most of us are being led, even in resiliency by our subconscious mind and this can be very dangerous. This very unconscious action is highly responsible for the hamster wheel that many of us experience in life and relationships. It causes us to run at the same things with different faces because we are blinded to the similarities of our past choices; blinded to our own mental and emotional wounds that cause us to feel comfort and even excitement from the very snake that has bitten us so many times before!

Part of cultivating a new you, is firstly, doing some serious work to understand the old you, the you that was created as a child and so forth up to this point. What made you? Do you pay attention to the things you feel and do you question where they come from? When we do this, it is like following a rope, and that rope leads to a box. Sometimes, this box is equivalent to Pandora's

Box FYI and it comes with an overwhelming amount of information if we are willing and ready to receive it. Your childhood upbringing and all of life's experiences up until now have in one way or the other played into who you are today and how you see the world. I am sure that you have known someone with a repetitious pattern say in relationships. They likely go for the same type of person or end up with the same outcomes over and over again. When we can remove our emotional attachment to our past "story," we can methodically assess our own patterns. Humans find comfort in the familiar, even if unhealthy. This causes us to repeat the similar situations over and over again. It doesn't start that way however; repetitious behaviors are often masked initially by shiny promise of "better" or "new and different." Take a relationship for instance, have you ever endured a painful breakup or walked away from someone with an understanding that it wasn't a healthy situation only to turn around and end up in a similar situation with someone "new?" It's because they are familiar and we haven't yet dug into our subconscious vulnerabilities.

Often times, the demons which we carry were given to us as children and if we wish to evolve in our own consciousness and experience a better quality of life and fulfillment, we will have to dig back into the beginning. I always recommend a good counselor whom you feel comfortable and connect well with. I put in many months of counseling after my split and the DUI. I didn't anticipate just how far back that I would need to go in order to move through the current obstacle which was my life at the time. I started therapy with the intention to understand why I had such an unhealthy pattern of men in my life, how to stop picking them, what I needed to change in myself and how to get my kids through this process with damage control. By the time I was ready to take a break from counseling, I had a whole new understanding of myself and my life in its entirety, and also the relationships and upbringing that set the patterns and my propensity towards toxic comfort.

While I was really open to delving into this mess that was me, I understand why so many people are not. It is uncomfortable and sometimes traumatic, but so is a future that's been destined to mirror the past. The definition of insanity is doing the same thing over and over again and expecting a different result. Many of us live in insanity, gravitating towards familiar patterns, and hopeful for a new outcome in an attempt to rewrite our past. We want to make the ugly stories pretty, so we tend to find similar situations and people to play "time machine" with. If we repeat a similar situation which has caused us pain in the past, and this time it goes well, then we "fixed it" and can live happily ever after right? Oh, so very wrong. This is how we end up on a hamster wheel of toxic relationships and situations. The most intelligent people can become victim to their subconscious preoccupations and I was no different from them all. While being an intuitive and fairly smart person in my own right, I had not understood that my repetitious choices signaled an area of myself that was very weak. Not only did I attract certain personalities, but I too received something from these disempowering situations. I saw the potential in people and fell in love with it. I believed that I could effectively empower or guide lost souls to realize and achieve their own greater potential.

I can't help but laugh as I write this part while realizing that my handicap relationally would also mirror what I now believe to be my greatest contribution to the world. On a worldly level, this need in me is fulfilling and full of purpose while on a relational level, it has sucked the very life out from my soul more times than I can even count. It reminds me of the movie Lord of the Rings, which to be perfectly honest, I have yet to sit completely through to this day. Anyways, I do remember the part where the person who held the ring often became negatively distorted by its power, a power that in the right hands and right situation was strong and used for good deeds. My unconscious handicap when directed consciously was equally my greatest power.

As I sat with my counselor and relived years of my life and experiences, I remembered an ongoing dialogue between my dad and I. My dad is a deeply spiritual man with a vividly creative and entrepreneurial mind; always has been. When I was a kid, he used to go on for hours in order to tell me about his different business ideas and creations which honestly went in one ear and out through the other. His enthusiasm was unmatched and while he didn't put these plans to action, he still creates and dreams larger than life. I owe my dad for a gift that I didn't even realize he was giving me, the gift of unwavering belief that this life is full of limitless possibility. I don't believe for one second that I would be where I am, or have created all that I have if it were not for all those hours spent absorbing his intense, childlike excitement for creating. He used to tell me that in God's eyes, I was "The Red Night." "Your soul is a deep red" he'd say, which I interpreted as being representative of intense power and great purpose. He was telling me that I was a warrior. Being talented as a child in writing and word expression, he would always tell me that I was "meant to be a writer" and that my words would "heal the hearts of mankind." I would roll my eyes and reiterate how I wouldn't want to be a writer because I only write when I feel something. I couldn't do it for a paycheck.

The process of getting into my early life experiences was necessary if I was going to understand my patterns of toxic partners and stand a chance at pursuing my future differently this time. I remembered things that I had long forgotten. My childhood years were really hard for a multitude of reasons but as I dug into this timeframe, I also had great appreciation for all that had made me. Every day after my counseling session, I would take a walk to the nearby boardwalk on the lake and think. I would think about that day's topic and sit with my emotions. As I sat there one particular afternoon, something quite interesting occurred to me. I had previously done a little bit of hypnotherapy training as well as life coaching and personal development, so I had a decent understanding of how the subconscious mind worked. I knew that this programming was relative to my current life patterns and that by digging in, it would allow me to better understand it. Once I understood my unconscious self, I could start to reprogram various areas of myself that were responsible for the patterns in my life, which had led to so much pain and brokenness. But as I thought about this early life programming, I made another connection. Like I said before, I owed my dad for the ambitious creator which I grew to be, as a result of which I was grateful and a deeper realization came to me in that moment.

My dad had spent my whole life telling me that I would be a writer and that my words would "heal the hearts of mankind." I was the "Red Night" he'd constantly remind me. As I sat on the dock thinking, having already started this book for my own healing, I realized that he had programmed me so early for the better. Despite all the pain in my early years, I was also built to pursue a powerful purpose in life. I was a writer because I had a deep desire to empower others. I wanted them to find purpose from their own pain, use it to end their suffering and run at their life. I knew what it felt like to be deeply hurt.

I wanted to help others heal their hearts and using my words was the only way I knew to do it. I was literally a writer, using my words to help heal the hearts of mankind just as my dad had said. And even though I had rejected both of those prophesies early on, here I am today attempting exactly that. In that moment, I realized that our subconscious isn't always unconsciously destructive, sometimes it holds our unrealized purpose. Before I got up to leave, a grin spread across my face as I also realized that still to this day, my favorite color just so happens to be red.

Chapter 4
Awaken

Let it die. Let it hurt. Let it go.

It had been a month and a half since I had moved out of my home and the divorce papers were well underway. I decided that I needed to get away. I needed a break from all that was going on in my life in order to pursue my healing process, and so I turned to nature. My best friend and I had planned a trip to Glacier National Park a couple weeks earlier to hike amongst God's glory and put to rest the mourning of my marriage. This was an ironic choice as my ex-husband and I had gone here yearly with our kids, and it was same place where he re-proposed to me, asking for a second chance after a previous seporation. It was our favorite place to be and here I was, planning to hike it 6weeks post move, as a spiritual closing of a major chapter of my life.

Not but a week before we were set to leave, the news reported about the recent fire outbreak that had broken out right smack in the middle of my favorite place on earth. Glacier was on fire, and the devastation was immense. Animals were displaced, residents lost their homes, many trees and trails were blackened and my much needed cathartic trip was not happening. I was genuinely devastated for the people, animals and businesses who were suffering due to this unfortunate event, but a different kind of pain also arose within me. My marriage was over, my family was forever changed and now my memories were being burnt in front of my eyes. It was as if God wasn't letting me go back there just yet. I checked the news daily only to see the updated numbers of lost acreage. I felt as though this devastating fire was personally set for me. What was, was no more and I would never return to what I had known before. If I one day returned, it would be changed as would I.

From the broken pieces, we are given ample opportunities to use them in the building of a stronger foundation. The foundation of a better life and greater living. If only it were so easy so as to sit back and know that this life would be a cake walk, where you would never be hurt, afraid or feel pain. But then, truly, what would be the point? Would we appreciate health if we were never ill? Could we know real love if we'd never experienced heartbreak? You see, it can be hard for our logical mind and even soul to understand that everything has a purpose. Everything given to us, good and bad, was given to use. The challenge is in the how.

At age 33, I found myself at my own rock bottom. With the ending of a very unhealthy marriage and my choice to leave, the decision to break up our family and put my kids through this unbearable pain while adding a second divorce to my resume was the hardest decision I've ever had to make. I didn't make it easily; in fact, I fought and questioned my own sanity for a good two years before making the call. Now, while I don't want to dwell on the ending of my marriage, I do know that it is necessary to reference this point in my life.

We cohabitated for 2 months before my place was ready for me to move out. During this time, I started counseling and started my healing process. I committed to my own healing because I realized that only I could heal me. Only I could restructure what had been broken. Ironically, when we break, we are never the same. We were never meant to be. We break so that we can put ourselves back together better, stronger and closer to our ultimate life's purpose. In order to do this, we must let things die, whether it's a relationship, addiction, memory or something else. Let it die! Hold it as a piece of your journey, but move forward. Let it hurt for a while. It's okay to get hurt. It's actually crucial to experience any loss to the fullest. We must truly mourn in order to move forward. Then, let it go. Accept the things that we cannot control and don't just make peace with them, use them as well. Use them as your building blocks to a great life.

I sincerely thank my ex for giving me the experiences necessary for me to become the person that I was meant to be. I healed to such an extent that instead of just seeing the hurt which had been inflicted upon me, I had empathy for what had made him. Empathy allows us to easily understand while not being a mercenary. We are not called to sacrifice our own healthy life for another. I can remember one specific counseling appointment where my therapist literally told me, "We get to choose who we give our compassion to." Say what? I had never seen compassion in that way ever before. Compassion, I thought was an understanding of another's circumstance or story which allowed us to give them grace for their downfalls, it turns out that is more like the definition of Empathy. Compassion is defined as the emotional response when perceiving suffering and involves an authentic desire to help. She was elaborating on this and conveying to me that if we do this for everyone, for everything that they do, we lose our own ability to set boundaries. My life had become quite exemplary of this imbalance.

What I had received from my partner was not compassion but instead, its complete lack of existence. I had first-hand experience with an individual who was completely incapable of compassionate reasoning or emotion, and I witnessed just how highly destructive it could be both for me and my children. I believe this also happens within the masses. A person may think clear and feel deep, but when presented with a story, circumstance or person through media per se, they are easily enticed to join the tone of the whole as opposed to taking time to use their own insight and compassion. That's why you will see a viral media post with thousands of narrow minded, insensitive and frankly mean comments and interjections. This shows a serious weakness of an individual, which is an inability to self-reflect on one's own values when motivated by the chaotic energy of a group. Why is this? How in a split second can so many people forget about integrity; forget their own values? Because it's unconscious. Unconsciousness will not build our character or grow us as people. Those who live life with a low consciousness are often less happy. They gossip more and complain often. They charge up their physiological body with negative and very real energy. This transpires into a negative overall demeanor all too often. When we relive our hardships and pains too frequently without a plan to grow from them, we are degrading our quality of life. Who or what is ever worth taking your happy life? Seriously! People really suck sometimes, I get it! Even more of a reason to steal back your power as opposed to giving it away. After my divorce, I was faced with multiple things that were completely out of my hands, things involving my children, my reputation and my life. These things can easily cause us to "see red" which I did so many times. The worst part though was that I never got the vindication that I solely desired from the offending party. The circumstances didn't change either. You know what did change? Me. My emotions were intense and my body was paying the price. My shoulders were like rocks, my sleep was terrible and my immune system was taking a beating. And all for what? A situation that couldn't be changed by a person who didn't give a damn. I became tortured while they slept soundly at night.

Learning to process things and let them go is a gift to yourself, not another. If nothing else, do it for the stubborn fact that this person or thing has already taken enough from you, the hell if you'll let them take any more! While some people will respond better to a peaceful acceptance and process of letting go, some of us actually need to get pissed. We need to get so pissed that we actually stand up for ourselves and powerfully decide to cut the line. We will no longer be at the mercy of that which we cannot control. Maybe this isn't a person. Maybe it's a fear. Are we afraid to face something because we are too riddled with fear?

Either way, we have the power to control how we respond and what we give our energy to. Make sure to set boundaries because they are a component of self-preservation, and self-preservation will protect you. Decide that you will become greater than your obstacles and that nobody deserves to degrade your destiny. Setting boundaries consists of limiting topics of conversation with unhealthy people or "keeping it business." Maybe it means speaking your mind at work to someone who had been disrespectful in a way that simply says "I won't accept being treated like that." Those are examples of boundaries and once they are set, we just follow through at which point there is little need to stress or dwell over the person or situation. We often get caught up because we feel it so personally. "How could they do that to me?" Or we get into petty details and insults in order to lift some of the anger off of our chest, thereby further escalating the conflict or situation. This only messes with your own peace and personal respect. No matter how "right" you may be in a situation, who really feels like their best self when they blow their top and just start spitting aggression? We feel much better about ourselves when we handle things calmly and collectively, and you can do this while still putting your foot down and enforcing your boundaries.

Personally, while I learned this early on in my Epic Year, it took so much devoted practice. I believe that it wasn't until about 8months after my divorce that I really started finding my stride once again. I set boundaries and worked hard so as not to make personal comments or petty jabs even if they felt very true and justified, so I simply enforced the agreements and boundaries. This became easier the more that I personally got healed. The farther that I got in my own healing journey and the more that I built my best life, the less that I truly cared about the decisions of the other party. Unless they were directly harming my children then it wasn't my concern, wasn't my stress also.

In order to pull this off, we must really grow in the self-awareness arena. We have to be able to look at our actions, emotions and behaviors in order to determine what outcome we are after. I paid close attention to my ego, my bitterness and my anger. I also paid attention to my intentions and often had to ask myself if it was going to support my desired outcome. An outcome that allowed me to have more peace and continue building the life that I felt was awarded to me.

They say that anger often comes from pain or a place of hurt. I would have to agree. When we are faced with an ending of a relationship, death or loss of any expectation, we are usually hurt. Maybe our heart is broken, our self-esteem is diminished or we are fearful and these feelings can often turn into outward anger, which can be anger that rises when in the presence of the offender or a similar situation. Pay major attention to those ropes of emotions that lead to your box of answers because they will give you all the insight necessary to grow, evolve and eventually heal the wounds.

In hindsight, I find it funny how egocentric we are as humans. No matter our age, we subconsciously believe that we are so evolved until father time comes around and schools us with experience, at which point we look back and see how naive we have been; prior to now of course as we again stand in ego! Just make a point on a regular basis to sit with someone much older than you, someone of another culture or belief system or someone who has lived through life experiences much different than your own. Practice calmly absorbing another person's perspective. When we do this, it helps train us to embrace a growth mindset which also humbles our perspectives and opens our minds to new angles and colors. The world gets awfully interesting when we get good at this. "Awakening" simply refers to the building of our own self-awareness and personal growth, and both are the tickets to a richer and much more fulfilled life as well as greater self-esteem.

If it no longer serves your growth, then let it die. Accepting the death of my marriage to someone whom I had considered my best friend for life was hard. But I am not only glad that I did it, I am glad that it was hard. I am grateful that I hurt through the process because it meant that I truly did care; that I had a heart, felt love and fought for it. I nearly fought to my own death, but I was proud that I had. I learned and I was hurt in the process of learning and I learned even more because I was hurt. The only thing left to do was to let it go. I had to let go when I had nothing to put in its place. Much unlike my previous pattern of going from relationship to relationship, this time I let go and mourned alone.

It is said that the most traumatic stress that one can experience is that of loss or death. Both loss of a life or of a relationship can produce a debilitating stress that's capable of making one feel as though there is no hope. The scariest emotion is that of hopelessness because without hope or faith, we have nothing to get back up off of the ground for, no reason to build a new life and no purpose. I can promise you that every person in this world has a purpose and that the most painful circumstances that we endure, were given to us in order to inspire or ignite our purpose. We firstly need acceptance of the unchangeable. Let it die, then mourn it. Mourn the loss of expectation or loss of love or person, and embrace your humanness enough to appreciate that which you can feel. You can actually feel because you are capable of love. Then lastly, let it go. Let yourself be totally free of constant torture and pain.

You have a job to do; a bigger calling in this life that needs who you've become. When a forest burns to the ground, we experience huge loss. We often use the word "devastation" to refer to the aftermath and loss of nature's beauty, but in due time, we start to see little plants and flowers sprout from the previously blackened ground. New seedlings show up where dead trees have fallen and life begins again from scratch. We are no different in as much as life will sometimes burn down everything that we once knew and leave us in complete devastation, just like the seedlings in the forest, from the ashes we will rise once again.

Chapter 5
Awaken.... Again

The interesting thing about deciding to evolve and grow above every other thing is that we often don't expect what comes next. I can guarantee you that you will be tested, maybe not right away but it will definitely happen, and sometimes you will nose dive! The most frustrating part of personal evolution is feeling like you have backpedaled or you seem to be too weak in a certain area to continue to move forward. That same damn wall just keeps showing up in your path, and you've got umpteen bloody noses and black eyes to prove that you haven't yet figured out how to scale it. At this point, it is crucial to tap into your self-awareness. Here within, lies the answers that will ultimately catapult you over that stubborn roadblock, but it is highly likely that the persistency of this obstacle is deep rooted. That means that you may have to dig harder and deeper than ever before in order to move through it. The good news is that once you do, you will see that path change, doors open up and endless possibilities enter your life. Finding ourselves knocked out coldly from the same thing that has intruded on our lives previously WILL happen at some point, and this is where we get to change our perspective in general. You know that big running wall at the end of the course on American Ninja Warrior? That's your obstacle. How many times you decide to run it is dependent on you and if you don't train, strengthen and grow, you likely don't stand a chance, but if you put in the time to learn it and anticipate the challenge when you run at it next time, you will get closer to the top or even scale it entirely.

    I have had many experiences with having to take on my obstacle especially, when I was in the intense points of my personal growth. It's so easy to plan our success and "know" how we will handle things before we are actually faced with them. Human beings are creatures of habit and familiarity is comfortable even when it sometimes seem to be destructive. It's easy to revert to our old comfort zones, but being comfortable isn't going to get us what we ultimately want. In order to achieve this, we have to do what likely feels hard and highly uncomfortable. Think about it; what great success was ever achieved by someone just chillin' with what they know. Staying right where we are won't get us anywhere great in life.

I remember the day that the first shift happened within me, it was early on and I had been carrying a lot of grief over my ex-husband's choices, carelessness about the ending of our marriage and lack of regard for our children's environment. We had been battling pretty hard over the kids being forced into a new "family" so soon with his new girlfriend, and I couldn't stop my fury over his disregard for what that was doing to them. I had no say; no control and our interactions were quite terrible. Profane text messages and venom on both of our ends flew like crazy and got us practically nowhere. He was going to do what he was going to do. That invoked a maddening level of rage within me, and I knew that my body wouldn't take it kindly for too long. I also knew that I had another 15 years of this! Then one day it just hit me. I was ready to be done caring about what I couldn't control within reason of course. It actually took rage and resentment for me to let some of the stress go away. I had spent so many years fighting with him to no avail, and wreaking my body in the process that all of the sudden I refused to let him wreak me outside of the marriage too, I was ready to put a stop to it. I hated the person that I would become while interacting with him. I had spent 6 years feeling like all of my flaws were actually who I was, thanks to him and while these tirades continued, so also did that feeling.

I had walked into my counseling appointment that day, and my counselor immediately looked at me with an inquisitive glance and said, "something's changed I see." She was right; while I actually wish that I could say this moment allowed me to carry on without discrepancy, it didn't, however, there was a major shift within me then. I had just had enough and so, I finally began to focus less on what I couldn't control and more on what I could. This consisted of getting my life built for me and my boys. I had 70 percent custody of my youngest, so that meant that I could control 70 percent of his home and do my best in order to facilitate his healing process, but the other 30 percent I would pray about. I knew he was safe with his dad, so I would pray for God to protect his heart when I couldn't, and be with him through all the changes. It's interesting when the fight yields less results than the calm. I figured this out about 3months after the split and started meditating on it. Every time that we had a situation and I remained levelheaded and collected, I would keep my integrity and thus, my view of myself improved and he could no longer get under my skin as he had in the past. I didn't do this because it was "right." I made the shift because I was done giving up my self-respect for anyone. It actually took until about the 9month mark for me to really put this into practice, but early on I knew what needed to be done.

I focused solely on strengthening my life and my business. That was the point where I started designing a life that I actually wanted to have for me and my boys. I had committed myself to staying single for me and for them until I had built the life that I always wanted to gift them with. This part of my journey gave me power in a situation where I felt so powerless. The power came with a sacrifice as most does, and I knew that I had to go through it alone for a while, possibly a long while. I knew that it would be so tempting to let someone in to share in my successes and frustrations, but I also knew that it wasn't time for that. I had to focus strictly on my goal.

Immediately following that grandiose plan, I had failed miserably. I got a little too close to a friend whose journey was destined to be far different than my own, and I lost sight of my focus for a month or two. I partied away too many of my kid free days which set me back a weeks' worth of health and productivity. I was not following the plan one bit. This became blaringly obvious when my main focuses all began to suffer. My business, health and patience with my boys all showed a major back pedal. So, there I was, picking up every piece of myself once again, dusting off my wounds and having the "can I get a do over" talk with God. I made hard cuts at that point and went back to semi-solitude to focus on finishing what I had set out to build; my life.

    I had felt so guilty for failing. I sat in my counselor's office enlightening her with my month of "oops" and explaining how I had just chosen to eliminate the obstacles which I had weakly allowed into my path. Again, my head hung in shame, so I didn't want to tell her about everything but in the name of transparency, I did anyway. I owe both her and my sister for bluntly telling me that I was looking at this experience all wrong. Not only did they not agree that I had failed, they both saw that I was tested and both agreed that I had passed. Really? I thought long and hard about this. A few years prior, I would not have possessed the ability to make others uncomfortable in the name of self-preservation, nor could I guarantee that I would have been able to sacrifice the very things that had become my own personal medication, but here I was doing it; the right thing. With this perspective, I was gifted another level of power within myself and it looked more like confidence. I now believed whole heartedly that I could do this; that I could change.

I believe that when we seek big change that leads to big reward, the process needs to make us stronger before it makes us new. That is why we are tested, so that we can navigate the same rough waters which have totally swallowed us so many times before, only to finally use a different strategy, one which gets us through the storm and beyond it. Now, we can sail and not because there will be no more storm but because now, we know how to navigate it.

Every test is an opportunity to either succeed or fail. With every failure is the opportunity to learn and when we know better, we can do better and eventually succeed. With each success, we grow stronger and more powerful because we believe in our own ability to create in this life. It allows us to grow from misfortunes and absolve ourselves of shame. What a gift! Once you realize just how powerful you are, there is nobody who can ever touch you. Your past cannot be dangled above your head when you have proven your growth with actions, and your future becomes much less daunting.

One impactful moment for me was during my sheriff's labor. I and a few other orange vested offenders were assigned to work at the landfill that day. As we stood on the side of the road clearing weeds that would surely grow back within a week in 90-degree heat, we chatted as we counted down the minutes until our day would end. Me being a life coach, decided to prompt some personal conversation. While I tried not to be intrusive, I was curious about everyone's story. How they got here, why they were here and where they saw themselves going in the future? I remember feeling very sad for them. One thing that many of them expressed was their fear of "being happy." When I casually asked them to elaborate on this, they openly shared that they were afraid to be happy out of fear of having it all taken away the next minute. They essentially had a strong fear of failure which felt almost inevitable to them. I couldn't personally relate to this because I knew that I would never be here again! I knew what I had done, and I was highly confident that I would never be here again. Not only that, I was confident that this experience was necessary in order to propel me into a better self than I had been before it.

As I tried to understand what it must feel like to believe that one is defective and incapable of dramatic transformation, I asked them what they desired for their future. Both individuals who had expressed their fear of happiness stated that they so much desired to finish college or enter a profession along the lines of counseling or public service of some sort. They desired to help others but interestingly enough, had little faith in their ability to help themselves. This fear held them hostage to the life they've known instead of allowing them to seek to become better. They merely tried to "stay out of trouble," and their life choices reflected that. Instead of completely ridding their life of substances, they relied on "less destructive" substances and instead of making hard cuts in their social environments, they merely kept the same company, but chose not to enter a public scene with them. I wondered if they were missing the "Why" or the "How." Did they not have a good enough reason to live a greater life or did they just not know how to go about it?

I still cannot answer this question for them. I do know that when we get hurt, finding motivation to run at life can be a rather difficult task. They were just surviving, and survival however isn't living. I believe that no person was created to only survive life. They were created to be great. The difference lies in how we view the negative experiences that we encounter.

With that said, no road to greatness is smooth sailing. Even after finding my footing from my divorce and DUI, I still had moments of failure. Moments where I didn't cope with my pain healthily or where I found myself struggling with the same handicaps which I had spent months trying to eradicate. But I didn't quit! I knew that scrapes and bruises were a part of the growth process. They were signs that I was being tested, and being tested allowed me to assess where there were still cracks in the foundation so to speak; the foundation of myself.

Simply wanting to change isn't enough. We must chase the person which we want to become. In fact, take "want" completely out of the sentence and replace it with "will." Chase the person that you WILL become! Expect that along this journey you will faulter, but refuse to simply survive. Refuse to merely avoid trouble or bad decisions and most importantly, learn from the consistencies which plague you. The harder they are to overcome, the deeper they are rooted so instead of giving up, become a digger. Grab a backhoe if that's what it takes to dig, and dig with the best of your ability. Doing that will get you somewhere this time, and once you see that you can overcome even the most smallest piece of the obstacle, you will find faith in yourself. Faith in your ability to not only survive, not just overcome but also your ability to rise far beyond the life which you have known.

I wrote much of this chapter during a major fail along my journey, and I found myself surviving for a couple of months entirely. It became evident to me that I was surviving by my state of my mind, business and choices. All of them reflected a person who was not at their best. I felt as though I was half assing it as a mom which was the worst feeling. My business wasn't getting the attention that it should, and it was clearly showing in its numbers. My body was NOT healthy despite a fairly clean diet and regular fitness routine. My anxiety was high and I was waking in the morning in a dark fog that I hadn't felt in years. There were details of my life and behavior that I was grateful nobody knew about. They made me a hypocrite and a failure. I wasn't healthy and I was not my best.

This wasn't the "plan." When I had left my marriage, I had decided to focus on 3 things; my kids, my personal healing and my business and for the first few months, although painful and hard, my commitment paid off. I wasn't a Rockstar at life by any means, but I was a fighter and my fight was blessed. My life with my kids was blessed, my body was healing emotionally and physically and my business was at the best it had ever been. During this time of focus, I still felt hurt. Sometimes, I would spend an hour or two crying after the kids went to bed or at nap time. The difference was that I let myself hurt, but I continued to do what needed to be done. I focused solely on the things that would bring me to a new level eventually. Even in the midst of extreme pain, I was making a better life even if it didn't feel like it in the moment. It wasn't until I crashed and burned a little that I was able to see the difference. I allowed myself to focus on my pain and not in a healthy way. I medicated and ran from my pain. I drank too much, too often and filled my empty voids with unhealthy thoughts, relationships and wasted time. I failed. Or so it felt.

I don't honestly feel it in my heart to judge others especially, those that harbor deep pain. I understand what it feels like to hurt deeply, and I know how tempting it can be to succumb to the temptations that promise even a second's relief. My hero, Tony Robbins always says that humans are hard wired to avoid pain and seek pleasure. Truth! Only this isn't always an easy thing to understand. Take drugs for instance; a person who uses any substance to avoid pain is only ridding it for a small moment. Ultimately that moment will end, and they will be left lower than where they started. Hence, how addiction happens. We have to keep medicating the very things that hurt us and ultimately hurt ourselves worse which plays into the vicious cycle of highs and lows and medicating. I get it. Pain sucks and that's putting it mildly.

I personally have always related to people who hurt deeply. I once described to my councilor that I felt that God had created me with the purpose of meeting those who are "in the trenches," people who are hurting. He assigned me the duty or gift of meeting those there. What it took me a whole 33 years to figure out was that my challenge in this would be pulling myself out of those trenches. Being someone who could lead people out, but not carry them and not staying in there too long. My setback was one of those moments where I hung out too long. The big guy himself had to come grab me. He had to remind me to put the oxygen mask on myself first, or

I was of no use to anyone. Heal myself and then lead others in their own healing.

So, now that I had found myself camping in the trenches, what was I going to do about it? I had two choices; shut the door and make myself at home or pack my bags and make a plan.

Luckily, I am a damn good planner and so the plan began. I cut unhealthy relationships and ties, cut overuse of any type of medication whether it be people, alcohol, food etc. and there I sat. I sat with my discomfort, my pain, my back-peddling pitfall just me and God.

My plan consisted of a full spectrum of self-awareness and personal healing, which I will cover more in the following chapters. What first started out so hard, had after 2 weeks become cloud 9 living! I had felt better than I had in probably my whole entire life and I was doing this completely alone. By alone, I mean that I had allowed in no companions and was relying solely on God for direct support. I was a single mom, running a business, supporting my two boys throughout my healing process and I was doing it so well!

About 2 weeks into this new awesome life, I woke up one morning with this strange bulge on the right side of my neck. I had noticed swelling in my neck on what I remembered being both sides about a month before, and couldn't figure out if this was sudden onset or if I was just noticing it now since my two weeks of self-nurturing had led to a decrease in weight and excellent decrease in swelling and inflammation. Being the holistic minded person which I am, I turned to Google, like any logical person would. After a couple of exuberant hours, I was half convinced that I was dying. I had some cancer or something, and this scary moment was going to last forever. This was a day where I did not have my boys and I spent the entire day spiraling into oblivion. I cried all day. I was scared and depressed contemplating how my boys would do without me or losing me. I was terrified to throw myself into the realm of allopathic medicine. The next day, I ended up making my first doctors' appointment in years. As I sat there with my 3year old, I waited for the doctor and wondered what she would say. How would she talk to me and would she see something terrible right there?! She came in and I was pleasantly pleased with her demeanor. She was nice and respectful of my knowledge and position in holistic health. She felt my neck and confirmed that she didn't feel a mass which she normally sees in cancerous situations. She said that she didn't feel that it could be anything of high concern, and offered a CT SCAN for my peace of mind. I declined for the time being and went home.

I spent countless more hours researching to no avail. The following day I chose to stop by our local natural foods store and speak to the in-house naturopath. Immediately, when I showed her my neck she asked if anyone has tested or looked at my heart. After I replied "no" she then asked how my blood pressure was, to which I replied, "abnormally high." She then gave me that "probably something to look into" kind of look, and proceeded to talk to me about my heart. I spent a couple minutes wondering if her reference to "my heart" was physical or spiritual. In a short second it became obvious that she was dually referencing both as one. My spiritual heart needed complete healing as well as the physical organ which also went by that name. I stepped back for a second because I had not shared with her about my divorce 5months prior or my recent decision to fully transform my healing.

I have never been someone who is comfortable being vulnerable. The thought of crying in front of almost anyone is mortifying to me. This likely goes back to my early years of self-soothing my emotional pain and anxiety as a child. Because I didn't know how to ask others to help me and because I didn't have answers for how they could help me, I could not break in front of them. I learned to do this alone, just myself and God.

And so, this woman continued. When confronted with uncomfortable emotional situations like this it is easier for me to talk in third person; as if myself is entirely someone else. I did this with her and as a matter of fact, I informed her that I had been through a divorce, was a single mom and after back peddling a little I bit had recently committed to a complete restoration plan. She vaguely continued to emphasize that I needed to "let God heal my whole heart now," however that may look. She walked me over to some supplements, and continued on about how sometimes God puts us in positions to speak for him and stated that basically that was what she felt she's being called to do for me. To speak for him to me.

At this point I broke. She broke me. This single mom, post-divorce, afraid for her life woman whom she felt compelled to reach that day. I cried a lot. My 3year old at the time was in my arms resting his head on my shoulder as it was close to nap time. I cried hard in the middle of the supplement section. She briefly put her arm around me in a slight hug, and then moved her hand up to the spot on my neck. She quietly started to pray for me and as she did this my son looked up, realized that I was crying and asked why I "had tears." I could not yet answer him and before I knew it, he had raised his little 3year old hand to the opposite side of my neck and held it there, simply looking at me and following in suit of this woman as she continued to pray, while I continued to cry.

I had nothing left, no pride and no more fight. I was vulnerable and I had no choice but to show it. Before I left, I deeply and sincerely thanked the woman. I walked through the store with no care that my face was obviously tear stained and plenty of people were likely aware of my emotional breakdown. I paid for our things and we left. Once we got home, I got my son in bed and sat and I deeply thought about what that experience was supposed to teach me.

I knew 5months prior that this was going to be a transformative time. An "EPIC YEAR." It was as if God had communicated to me somehow that he intended a major transformation for me and I had known all along that this would likely come in the form of many hard experiences. But "how hard" I wondered? Was something wrong with my body or was this a scare meant to teach me something else? Or both?

I identified two things that day; that I had a big problem with vulnerability and letting people in and that I carried way too much fear in my soul. I had been carrying a heavy load of fear for my whole life and somehow, I knew that God wanted me to heal that. While this insight had occurred to me, I still went on for a few days to suffer in debilitating fear and depression off and on. After a few days of this my mom called. I hadn't told anyone about this prophetic woman who broke me into a vulnerable mess a few days prior. I decided to fill my mother in a little and again, reluctantly, I cried.

She said "honey, this is crippling you." She was right. I was paralyzed by my fear of the unknown. Prior to this, I was doing so good! I was on cloud nine, healing and thriving! I felt unstoppable, happy and deeply grateful for everything in my life, even the painful things that had led up to this point. So, how was I now so low? The answer was fear. No matter the outcome of my situation, I needed to dig into my deeply rooted fear of the unknown. The place that held all of the things which

I could not control nor predict. The place where the truest form of faith lies.

Being in holistic health, conventional medicine just wasn't a comfort zone for me. I had terrible experiences with allopathic medicine and physicians in the past, and avoided it at all cost. Cancer and those sick with it were wildly hard for me to face as I absorbed the suffering of others, and the environment exacerbated my anxiety to the fullest extent. A lot of fear accompanied this topic for me. The thought of being around sick individuals who may pass away as well as the hospital setting was just too much for me to handle. I feared that this would now be me and the thought alone reduced me to nothing. It broke me yet again. I tried to fake a confident woman and mother, but I just didn't have it in me and so, I shut down and tucked myself away from the world again.

A couple months prior to this, we lost my grandma and just weeks before that my mother lost one of her closest friends to colon cancer at only fifty. Death was in the air and I was inhaling it so deeply that I almost wanted to stop breathing completely. These circumstances all represented fear and lack of control. Both were things that sent me into a tailspin. I wanted something to take away the anxiety, but nothing would.

I realized then that there were limitations to my faith. I had full faith within certain parameters which I had drawn myself. I trusted life's plan for me "if" it was within a certain spectrum and life-threatening illness or death were not on the list that I had made! The list of what I would and could accept. My ability to look at my own emotions and follow my ropes to my boxes is what allowed me to come to the realization that I was trying to control life too much. I thought that if I did X, Y, Z that I could control my outcomes and direct my life in a way, one which would only challenge me within a certain self-decided degree. Ha! Life gives you what you need and makes you into the person it needs you to become, and not the other way around. We are merely the clay working and learning in the molds that have been given to us.

I was highly disappointed in myself. I had started really strong in my intentions to rise from the ashes, and yet once again I lay flat on my face. Shortly after my 5th damn near nervous breakdown in only 6months, I clearly understood that I was not being only molded but tested, and as I lay face down in the metaphorical mud that was my life, I decided once again that it was time to stand back up.

I embraced the fight as well as the failure knowing fully well that God obviously had a big purpose for me. I wasn't weak indefinitely; I was just being transformed through failure and fight. I believed at that point that he wanted me to keep fighting; to keep standing back up and when I embraced that concept, my perception of myself started to change as well. I had ridden a rollercoaster of self-esteem. High when I was handling things well, and so low when I faltered. My self-worth was contingent and therefore fleeting. I had made the mistake of seeing this journey as an ugly reflection of who I was instead of an intricately designed process purposely molding me into the person I was supposed to become. We are not our mistakes. Our value is in our purpose, and no matter how hard or how often we may fall, it is our job to keep getting back up and holding onto our determination, so that we may see our greatest potential into fruition.

Chapter 6
Your True Value.

"And just when you thought your day was going incredibly well, hardly anyone responds to your Facebook post and the rest of the day is shit."

Did you "LOL" or cringe at that statement? As if finding self-worth isn't hard enough, now we add the constant fluctuation of outside affirmation to the mix. With social media, we have now become a society who creates an "identity" in the form of a profile for the world to see. Maybe humor is your forte or maybe its sex? Either way, our dependence on attention has grown rapidly and is fed or denied to us minute by minute. While social media isn't the only unhealthy way in which we strive to obtain personal value, it is a substantial one and worthy of addressing. Much like a drug, superficial gratification is short lived. It lifts us and then drops us on our face all the while, thereby keeping us coming back for the next "high." Humans have a need for significance and will seek it in any way possible. Maybe it's a relationship, job title or outside praise. While this is human nature, it is more important than ever that we learn to ground ourselves enough in order to develop a true value in our own light. By finding our significance in the world around us, we hand over our personal confidence and happiness to the rollercoaster that is in this world.

Like I stated in the beginning, we were all designed to become great, and each individual's "greatness" is not for anyone to judge or award as we have no idea of the ripple effect that they were created to set off in this life. One thing is for sure, our individual greatness does not come from others opinions of how "great" we are. It is not in our superficial profiles and pictures, but rather in the value within ourselves. When we truly find value in who we are, we can then align with our purpose in this life and therefore bring greater value to the world around us. When we merely look for our value in the opinions of others, we let everyone outside of ourselves set our limitations. This is not only dangerous, but incredibly self-destructive.

Relationships are one way that we hand over our personal worth to another, often someone who hasn't even known us long enough to have a say in the first place! I discovered this on a deep level about 7months after my divorce. I wasn't really dating as I was focused on my kids, my business and my personal healing, but I would still occasionally meet men or they would seek me out through certain mutual connections. Because I wasn't ready to seriously date, I wouldn't entertain many of them and the ones who I did, I did so with a "just friends" attitude towards them. Even still, I was ill prepared for the single world. So many people out there are only looking to rush into relationships or simply medicate with meaningless sex. While I don't judge what another chooses for their lifestyle, I couldn't help feeling really devalued at times. Not that I had even let these people close to me, but even in conversation I could tell that they didn't give a damn about where I'd been, who I really was or my ambitions for philanthropy work, they simply wanted and even needed to fulfill their own voids.

I had spent 6 years being devalued by who I believed to be my life partner. For 6 solid years, I was reminded that my value was superficial and that I was only significant if I provided for my partner's needs. If he couldn't extract from me then surely, he could replace me with someone else who could "fill my position." He eventually went on to fulfill that prophecy, and had a semi live in girlfriend 4 weeks after helping to move mine and my boy's things away from our family home. He didn't shed a single tear throughout the process prior to me moving or after. I was simply a person filling a position, filling a need; and when that time was up, I was nothing.

So, being around so many men who were only determined to fill their personal voids with fast relationships or cheap sex, just brought me back to that place of worthlessness. But why? I had to stop and ask myself where this was actually coming from. I hadn't given them anything to use, so why was I allowing their intentions to lessen my value? After some time of meditating on that one, the answer came to me. I was allowing them to assign me value in the first place. I was waiting, even as "just friends" for someone else to appreciate who I was, the way my heart worked and all the good qualities that I possessed. I was waiting for them to appreciate what my former partner never did. When their acknowledgement was shallow, I allowed that to communicate that my deeper gifts weren't significant enough.

WTF? Seriously though. I couldn't blame these men! Sure, they were kind of immature or in a different place in their life's journey than myself, but they couldn't touch me! They couldn't affect my self-esteem, my worth, my strength, my character or my happiness! The only way that they could get into those crevices of my mind was if I let them in. The only way another person can rob you of your self-worth is if you give it up to them. Stupid! Really stupid and really easy to do. We start this way of unconscious self-sabotage early on in school. We allow others to rate us or categorize us in clicks or groups. We value ourselves based on our looks, clothing or finances all the while and paying little attention to the greatness that we possess and the power that lies within our own control. Power in the form of personal value is capable of moving mountains and setting the world on fire! People that tap into their true value become givers not takers. They don't need to extract from others because what they give to this world provides an unimaginable level of personal value.

Imagine that you meet someone who you really like, and after some interaction they reject you. Maybe they stood you up for a date or said something cruel to a mutual friend, you'd probably feel the self-esteem hit from that one, right? Now picture this; you decide to take a 2week trip to Africa to a small tribal area where the population, including children and babies are getting sick and even dying due to poorly filtered water sources. The group you are going with is there to remedy this obstacle by not only installing water purification systems, but also training an onsite local medical team. While you are there, you volunteer to work with the sick children, nursing the strong ones back to health. You get to witness the gratitude from their parents as these kids begin to put on healthy weight and start smiling again. In your down time, you play soccer with the local children and are overwhelmed by the hospitality of the village. By the time you leave, many children have recovered and the light which you have given to these people is so appreciated that you can't help but feel the warmth and happiness reflecting out of them and into you. They dogpile on you for hugs before you get in the car, and a simple "goodbye" is full of tears from you and them.

Now, you get off the plane at home, heart full of purpose and run into that same person who stood you up or made you feel so devalued before you left. Do you feel the same? Can this person even begin to touch that which you have become throughout your recent experience? Can you believe that you even liked this person!

Funny how it works isn't it? <u>THAT</u> sweets, is "True value." True value is birthed from the intangible things in this world and what we become throughout the entire process. It does not come from photo "likes" or attention nor should the power over it ever be put in the hands of another human. True personal value is what lays the foundation for a life well lived. It guides us towards our purpose, heals old wounds and gives peace to our soul. We get to assign our value in this life, and the trick is to do it with conscious personal growth. If we simply tell ourselves how wonderful we are, but do not seek to live in a way which contributes value to others than we cross the line of confidence into ego, and the ego is selfish. The ego takes from others, justifies it and believes that it deserves whatever it wants. The ego doesn't self-reflect or apologize. It is very immature and this immaturity does a hefty disservice to our personal growth. One reason why I love the process of personal improvement and growth is that it allows us to shed past mistakes and shame while creating a better, healthier path for the future. We can't change where we have been or anything that has happened prior to now, but through growth we can be better so that we can do better and that is uplifting.

Sit down and ask yourself what you appreciate about who you are today as well as what areas you'd like to grow or change? Appreciate your body where it is and commit to nurture it into its best self because you value it greatly. Get your heart right and most importantly start giving in some way. Find your true value and refuse to let anything other than your choices define that value. Stop assigning validity of your worth to others. Not people, ads, media, status, money or any other material or superficial source, this includes relationships. If we find our true value first, our expectations in relationships change. Not only are we better at giving and contributing to other humans, but we have a higher standard for how we allow another to treat us. This doesn't just pertain to romantic relationships, it may be friends, coworkers or even family. We teach people how they are allowed to treat us by what we tolerate and the boundaries that we do or don't set.

In my single life frustrations, I had realized that not only should I not assign my personal value to another, but I should be holding others to the same standards of respect and character that I held for myself. That was not only my right, it was also within my power. I get to choose what kind of people are allowed close to me, and if someone didn't meet that standard of respect then I hold the power to protect my value by drawing hard boundaries, and that's the end. No need to dwell on it, just make the cut! Once you learn this skill of self-preservation, you will be highly impressed with the caliber of partners, friends and relationships that start entering into your life. We tend to sometimes believe that good people just don't exist, but that is not the case here. The problem isn't them, it's you. My environment of disappointment was all me. It was me making excuses or allowing people to hang around who honestly didn't deserve to make the cut. We can hold compassion for others, but it should never be to our own detriment. If you choose to fight through the hard times in life and persevere through growth and humility, then why the hell should you carry those who choose not to fight for themselves? You will wreak yourself allowing destructive energy like that into your environment.

While everyone has a different pace along this journey, some people get too much superficial gratification being a victim, and may never choose to rise above this. By carrying them, you will only wear yourself down and diminish the light which you have to gift the rest of the world. I struggled with this piece of the process because I truly believe that every person is capable of any level of change, however, I had to acknowledge that it was not my job to bring them to their own realization with feet dragging. We are created to lead, empower and inspire. We should absolutely not ignore those who are struggling just as we do not want to be judged or ignored during our times of suffering, so learning to identify the difference between someone who is struggling to stand up and someone who has no desire to even try is very critical.

As we develop a greater personal value, we will also find greater value in those around us. A year prior to my divorce, I had set it in my heart that one day I would work with or meet Tony Robbins. I didn't have a whole lot of mentors in my close life and had latched onto Tony for his incredible heart, exceptional business skills and powerful impact on so much of the world population. I dreamed about doing one of his more intimate, smaller group business trainings which at the time was not something that I could afford. Fast forward a year later and there I was, just divorced, single mom and had hit the best month in my business's history. The first thing that I did was invest in training with Tony. I prepped and planned for 5months before the event. It took me 14 hours to fly to Florida as my air miles afforded me 3 different layovers.

When I arrived in Florida, I got my second wind. I barely slept for two days prior to our first training day and when I walked into the convention center, I immediately found my way to the bathroom, hiding in the stalls to try and control my emotion and even tears. I wasn't just emotional because I was finally getting within arm's reach of my hero, it was more than that. Throughout my divorce and devastation, I had grown my business substantially, knowing that I would be supporting my boys on my own. I was fighting tears of gratitude and awe of this life which was finally starting an upward trajectory.

Throughout 5 intensive 14hr days in this conference, I had the privilege of connecting with so many inspiring individuals. These people all had businesses as well as personal stories of success, failures and fears. All of us were in one room with one mission; to grow. Grow our businesses, our networks, our mindset and skills. This powerful energy was unlike any environment I had ever been in, and I valued it so much. I truly valued these people because I valued these things in myself first. I went to this conference with my ambitions of helping people, philanthropy and all the emotions that accompanied a hard 7 months, and I met others with similar and beautiful energy. I carried on so many meaningful conversations and felt genuine support from total strangers. I also gave genuinely to people which I had never even met before. We were more powerful together because we were growing in our own value and it was flooding out of us and onto each other. This was wildly different than many of the relationships I had become accustomed to. The unreliable kind that might crumble if something better comes along or that are built on extraction and competition instead of being built on mutual altruism. The people that I met were truly interested in who I was as a person, and I felt the exact same for them.

One of the activities that we had to do consisted of mentally designing the person which we were creating ourselves to become. Tony is really good at empowering people. Being in the room with that man connects you to your own personal power, and I was definitely blessed to be experiencing it. The business leader, take life by the balls, philanthropic powerhouse that I was envisioning myself to become needed a name. This was the woman I was born to be. The woman who had purpose, who had risen from the ashes, determined to empower others to do the same. I thought for a moment and knew exactly what that name was. I wrote down "Red."

Before we left the event, Tony had us write a letter to our future self. My letter was written by a broken woman, finding her footing and just starting the climb upward to a greater destination. I knew that I had come a long way in the previous month, but I still had a way to go facing my fears, confronting demons, healing and building. Building not only my business, but my life, for me and my boys. The life that I wanted; the life which I needed. We then turned in our letters with no information on when they would be received. When I arrived home, I knew that I would have to covet this new perspective which I had attained, one where I sought out to put healthy individuals in my space and distance myself away from people and situations which would not support my direction. I also knew that like many times before, this conviction would be tested. I was no longer in my safe place; I was back in the real world with real challenges. While setting standards for self-care isn't easy, it is absolutely worth it. It sometimes means more nights alone and less social opportunities, but ultimately it comes down to quality over quantity. If we continually lessen our standard due to impatience, then we never get the ample opportunity to experience the quality of life that we ultimately desire. A healthy life aligned with people and environments designed to guide us to a greater purpose. That purpose is worth turning down cheap dates, unhealthy situations and consciously building a more empowering circle.

Through purpose, we become great and through contribution, purpose is often realized. When we give of ourselves in some way, we gain more personal value from the experience than from any other type of accomplishment or success. Tony Robbins says that "We don't give in order to receive, we give because of who we become through the process." There are two huge steps that you can take to grow your personal value substantially in a quick period of time. The first step is detaching yourself from other people's opinion of you. To really gain momentum in this area, I recommend simultaneously seeking out a healthy circle of individuals and relationships who align with the direction which you are headed and who believe in one's ability to create their own best life. This will make the simple opinions of others miniscule in comparison and also support your personal growth. The second action that we can seek to quickly enhance our personal value and view of ourselves is through giving. When we volunteer to help others or give of ourselves selflessly, we become a better, greater version of who we are, and this evolution guides us closer to our purpose in this life. Like the village children scenario, we addressed that gratification from selfless contribution will help put things in perspective and likely diminish our consistent need to gain worthiness from other people's opinions and affirmations.

Personal value will convey as confidence, and not false confidence. False confidence is conceded and over-explanatory. This is because false confidence is a show that's meant to mask deep rooted insecurity. Those who are falsely confident will easily tell you how great they are or list their array of accomplishments. They may put a large overemphasis on physical looks and superficial possessions. Not all insecure people are conceded, some are just so fearful that they won't be accepted that they try and help to "paint" you a picture of them in hopes that you won't see them the same way which they truly see themselves. Insecurity can cause us to work really hard to show others that we are special, whereas confidence is more concerned with making others feel special. We all struggle with insecurity, especially in this modern world, but self-awareness is the first part of transitioning out of false confidence and finding our true value. Learn to listen when others speak and complement them without speaking of yourself. These behaviors become habits and habits take practice to make as well as practice to break. If this is you, do know that it doesn't make you a bad person, it is just a sign of an area that you personally need to heal and grow. If this is you, then feel proud that you are secure enough to recognize it and desire to change it. Congrats, that takes confidence!

Growing in your true value is an essential component to designing your ultimate life. Once you've grown your value and opened your mind to a purpose driven life, you will start to understand why your life has gifted you both the blessings and trials. You were designed to be great and you are so much more powerful than you even know! Now, life will make you prove that power but not to anyone else, it will make you step up to a new level and prove your own value and power to yourself. It will bring out deep rooted fears, like a horror house or maze designed to be unpredictable, but that ultimately leads to a greater destination.

The catch is that the only way out is through, as you will have to face some of your greatest fears head on, alone.

Chapter 7
Running At Fear.

I vividly recall one specific day back in September. My divorce had just been finalized and I was taking my 7year old son on our weekly date. This week we chose frozen yogurt and a hike around the lake. As we walked together, I tried my best to absorb the moment with him in spite of where my head was. At the time, it hadn't been even two months of being a single mom and my ex had just informed me that his new girlfriend had just officially met our son and would be staying with them in my former home. I struggled with the fact that the "right" road which I had chosen to take was currently feeling heavy and sacrificial. I believed that all of these painful experiences were meant to teach me, but too often the minutes seemed to tick by slowly and they left me wondering why I had gotten the rockier path in the deal. If I was making the responsible choice, why did I feel so defeated? Why was I the only one mourning my broken family, and why was my son's environment and emotional well-being so out of my hands? I hated that my youngest child now also had to experience a broken home, was being forced into a "new family" within weeks, and I was helpless to protect him. It's one thing for life to school me with hard knocks but my kids!

As I struggled to put the unchangeable out of my mind and absorb the present time with my oldest kiddo, something happened. The hike we were on looped around a small mountain surrounded by the lake. As we walked, my son chatted away and frequently grabbed onto my arm out of fear that I would get too close to the edge and fall. I laughed to myself partially at the irony of my clumsy 7year old worrying about my footing, and partially out of appreciation for his adornment. As he jabbered, he said something so poignant that my hair stood on end, and I could feel God asserting himself through my son's words. He said "remember mom, if you fall off the edge, always trust the branch. That's what God tells us to do." Even though I was fully aware of his very literal intention behind this advice, my eyes welled up immediately and I hoped that my sunglasses would hide my tears from oncoming hikers as well as my little boy.

He was right. I had slipped off of the cliff and was hanging by a branch, clenching onto it for dear life, but that branch was rooted deep and it wasn't going to let go of me as long as I didn't let go of it. "Trust the branch mom" he had said. Even though I couldn't see the outcome, even though I was dangling off of a cliff, unsure and afraid, God had me. Fear made me question the outcome, the purpose behind the uncontrollable, but faith told me that everything was going according to plan. Faith said "You are doing it. Keep going. It will make sense soon enough." And so, I did.

Six months later, I sat on hold with the doctor's office for the fourth time in 3 days. I wanted my results. I had finally decided to face my fear head on and had started the process of figuring out the neck situation. Neck stuff can be really scary and I promise you that if you google it as I had done, you will surely believe that you are dying as I was starting to believe myself. The hellish part of the wait was where my mind went. I worried about being sick and a single mom. My heart broke for my kids having to watch me suffer or God forbid lose me at such young ages and one thing was for sure, there was no way that I could be sick in front of my ex. Anybody but him. I am not a person who easily asked for help and I sure as hell didn't ask for sympathy or attention, but had I needed it from him ever, it was replaced by an almost unphased coldness or a communication that I was pathetic or weak. I had lived like that for so many years and I couldn't feel that again, not now. Not after all of these months of fighting to rebuild my self-esteem and repair my sanity. I could handle a lot, but I couldn't handle that.

The blood-work that they had done on me came out fine, but now I awaited the ultrasound results. For the fourth time in three days they informed me, yet again that they had the documents, but the provider had not reviewed them yet. I hung up discouraged. I was actually beyond discouraged; I was losing it. My fate was literally at the hands of a practitioner who just hadn't had time to get around to letting me in on it yet. I couldn't take it anymore! I threw myself on my bed and cried, hard. I prayed hard too. I usually didn't make a habit of asking God for anything with exception to healthy kids. I usually only prayed in gratitude, but this was different this time. I was literally on my knees. I just remember crying out loud, "God just let me stand up!" I had accepted everything he had given me. I had picked myself up so many times just to fall yet again. "I am ready to stand up, please just let me stand all the way up!"

I had big plans. All the painful events of the previous seven months had birthed a new, stronger and more determined me. I was gaining my strength back, and then at some point I was ready to run at life, if I wasn't dying that was. In the back of my head existed the fear that there may be something seriously wrong with me and it was holding me back. Fear has a funny way of doing that.

The phone rang. The Dr's office was calling to inform me that no mass or formation was found, and that the ultrasound was quite normal, but that they wanted to see me again to determine what other actions may be looked into. While this wasn't a clean bill of health and I wasn't totally in the clear, it was good enough for a breather! So, I decided to take a break from fear and embrace the unknown. While I wasn't completely at ease, I knew that my results could have yielded much worse, but that obviously wasn't in the plan for me right now and for that reason I was very grateful. The unknown had tortured me throughout my entire life, from fearing that special school in 2nd grade to this mysterious neck swelling. It debilitated me with fear. The irony in all of this was the symbolism. Here I carried so much fear of the unknown and yet my life at the moment was completely unpredictable. When I decided to leave my marriage, I did so with absolutely no idea if I would be able to drive my children or pay my rent, not to mention being single for the first time in 14years and now with kids! Yet, I did it, and I did it in faith. So, what was it that made that leap less scary than the situations that I was currently facing? The answer was pain. After a certain amount of pain, we will leap mountains in order to find relief but it takes a push. Pain pushes you past your threshold and into growth. Growth teaches us that we can evolve and even concur fear. There is a saying that goes, "Figure out what scares you most and do that." If we chase what scares us, we concur the very thing that we were once a prisoner to.

From the time that I was a child, I have had a similar reoccurring nightmare. While it differed slightly, it always had to do with a demonic creature in a dark house which was after me. I was terrified and often couldn't see it or predict where or when it may get me. I couldn't anticipate it. I was a sitting duck. In the early years of this nightmare, it would pop out at me and I would then wake up. I don't know when the turning point came for me, but one faithful day the dream changed. I still continue to have this dream only it ends differently now. The narrative has changed. It still starts out the same, an ominous presence accompanied by intense terror that I can't yet see with my eyes. I have no idea when or where it will get me, but I know its lurking. The difference now is that while it starts in fear, the fear transitions into a fire, an energy, a fight. I find myself turning around and charging at the illusive demon with all I have and while in mid run, I see it. I keep running and just before I reach it, I wake up.

I thought long and hard about this dream because in a metaphorical sense it represented my life. Concurring fear, rising up and deciding that the only thing worse than running at what scares us, is just sitting and waiting for it to get us first. So, for the first time in my actual life I decided to attack my fear. I planned to put myself in situations which normally I would not be able to do. I dove into the thought of death. What if I had a terminal illness? What would I do and how would I live out my time? How would I make peace with leaving my children and loved ones behind, and the process of evolving into the next life? As dark as this had originally felt to me, the more I approached it, the lighter it became. Part of this process of my life brought me to the realization that we all need to dig into what terrifies us, and I needed to face a brutal fear of my own; mortality.

It was at this moment that I decided that fear had consumed too much of my life and I was done succumbing to it. I may not be able to control the demons in the room, but I was done hiding from them. I was ready to run and so I made a plan; a plan to run at fear head on, a plan to grow. I needed to finish building myself mentally, emotionally, physically and spiritually. I had done a lot of counseling and had a good logic moving forward. I was already making hard cuts and dodging prospective relationships which did not align with my new standards. My mental self was gaining momentum and my emotional self was following in suit. As I made healthier conscious decisions and sacrifices, I gave my heart a chance to heal and therefore my emotional well-being started improving too. The two areas that still needed work were my spiritual self as well as physical. I was tired though. I had turned 34 at about the 10month mark of this epic year and along with the upward momentum, came the need for even more endurance. The previous months had already taken me to the brink of what I thought I could handle and knowing fully well that I still had more ahead of me scared me the most. I celebrated my birthday with a tired realization. My understanding was that fear and I had known each other all too well in that thirty-four years, and the one thing that the last 10months had taught me was that when you need to grow, you will be faced with something that scares you real bad. I had experienced more fear in this 10month timeframe than in most decades of my life and not ironically, I had also experienced the greatest success and spiritual growth. As I healed, I wrote, as I fought, I wrote and as I learned, I wrote more. As I stood on my last leg of my epic year, I could feel the heat turning up like the last miles of a marathon, having to push harder when your body is at its limits. The mind will now take you along the rest of the way and faith in the purpose will trump fear. So, I pressed on in faith and sweat.

While I was into fitness and ate pretty clean, my stress had been hard on my body. While growing up, I was really overweight and while I liked sports, my cardiovascular abilities were limited. I was the last in the mile run, basketball wore me out and agility was not my gift. I really had to push myself as I got older and into fitness. I still didn't consider myself an athlete, but had become rather athletic, on my own time that was but I was not comfortable having another person push me and had avoided it at all costs.

Part of my new plan to concur my deep-rooted fears was to push my body or rather my mind. My mind would panic a lot, thereby hindering my body and setting its limitations. I needed to push my body past my minds previous limits to gain the strength that I was after, but I knew that I would need assistance. So, I decided to hire a trainer. I would actually seek out someone to kick my ass, push me and I would pay them to do it. I asked around and hired one of the most hardcore female trainers in the area. I had been seeking someone to help me build a stronger body that could match the warrior I was creating on the inside and when we met, I knew that she was my girl. This chick was going to kick my ass, but since the previous 10months already had, I was ready to have some visible muscle to show for it.

There were days during my training that I felt sure that my muscles would explode, or times where I was positive that I wouldn't be able to pull off another rep let alone 7! I was so sure of my limits, but she pushed me further. She made me keep going and in return helped show me my body's real capabilities. It may sound silly to find so much metaphorical symbolism in weight training and fitness, but it really is eye opening. I realized that I had consciously chosen to grow in multiple areas of my life throughout this year and in each area, I sought out someone to push me further down the lane. In order to grow from my pattern of unhealthy relationships, I found an excellent counselor. In order to evolve my business, I worked with a coach and invested in a major, life changing business conference. Now with my trainer, I was focusing on my body and subsequently, my mind.

While it was essential for me to elicit support in these areas in order to achieve real progress, it would have all been useless had I not decided to embrace the challenge that faced me. I opened my mind to others perspectives and expertise as well as their experience. I sought these professionals out willingly because I wanted to grow, I wanted to do better, be better and achieve my goals. I was designing my own dream life. I knew that it would not be easy, nothing that is worth anything ever comes easy, but it was worth it to me, so I was willing to work for it.

Through all the hard work I saw change; I created it. Nobody gave this new life to me. Not my perspective, my mindset, my body or my success. I had chosen to do what was hard and knew that I needed to in order to reap the reward of a better life and healthier self. While I pondered to myself one evening, I wondered why everyone doesn't do this. I mean, it's an ass kicking but ultimately isn't your dream life worth it? Now, I can't answer for anyone else but I can tell you that I thought extensively about that question and in that moment, I was re-routed back to the conversation with my fellow sheriff's labor crew the previous summer. I remembered the young man saying that he was "afraid to be happy" for fear that it would be fleeting. After all, everything is truly fleeting right? So, why work for it if it will just go away? I clearly understood that. It feels much less intimidating to stay in a place of familiarity. We subconsciously believe that we already know what will happen and in return, life doesn't feel so scary. The truth is, while everything is fleeting, it isn't necessarily temporary in a negative form. Sometimes, we learn and sometimes we grow and surpass our previous situation of life. We actually grow out of it because we are ready for a new level of being, in which case "fleeting" isn't the right perspective. It is as it should be and with the tides comes opportunity. It's like riding a wave, you just jump up and whether you surf for miles or crash immediately, you will have learned more for next time and when the next wave comes, you will try again. That is life!

Embracing the fight equally as much as the success because they are always interconnected. There is no success without failure, however, if you stay mentally stuck in a belief that you are helpless to your circumstances, then you are a sitting duck while waiting for the wave to crash over your head again and again.

Growth isn't scary, death feels scary. Even still, those faced with their imminent mortality often grow a new perspective and accept the life after death scenario as growing from this life. If we can envision all of life and its twists and turns as growth, then they aren't so scary. They instead become a method of evolving from what we once were. Once you get this concept, you will have the opportunity to run at fear by essentially running at life.

In a video script which I had written for my coach training school, it says "The mind is a powerful tool. It can hold us back from our greatest passions, or skyrocket us into achieving our greatest dreams." I was done letting anything or anyone hold me back from my dream life of peace, purpose and love. It was now time to run. As I started my metaphorical sprint, I realized something.

I realized that in order to "run at my fears" I merely needed to let go of them. I had been so used to the fight which the previous months had taken that I hadn't completely understood the power of letting it be. In this moment, I dropped my weapons and slowed my pace while realizing that choosing to dismiss emotions which did not serve me was more powerful than anything I had done thus far. While the sprint is necessary at times, it is meant as a momentary action of survival. The real power for long term wellbeing actually resides in one's own ability to separate themselves from fear and in that moment, I stopped running, sat down and started to meditate.

Chapter 8
Purpose From Pain

I fully believe that we are all individually called for a greater purpose and that our own unique greatness is essential to this world. When God or Life, whatever higher power you believe in decides to call on you to use this greatness, it will first have to train you for the mission ahead of you. This training will come in the form of hard days and experiences as well as personal realizations, such as the luxury of selflessness and pure gratitude. In order for you to transform, your soul must be allowed to come to life.

A heavy soul carries many emotions such as shame, guilt, anger and resentment and these emotions will not serve you positively. They are the unconscious leftovers from painful experiences and they are meant to make you miserable until you finally use the initial events as a tool for growth. If you touch something hot long enough, it will eventually sear your skin, thereby making it uncomfortable enough for you to remove your hand. The body has designed this pain mechanism to protect us and keep us healthy. The emotional body is very similar. It will make you severely uneasy until you tend to the wound and move forward. This isn't a gift to others; it is a gift to you. You are giving yourself peace and in the long run, rising to a new level of conscious maturity.

Now, don't get me wrong, I believe that it is still very important to verbalize pain and let out anger and negative emotions. Sometimes, we deserve to be pissed! But once it's out, don't let it back in by constantly reliving the story. I personally struggled with that one. I relived much of my life and my stories and in return, I found myself more and more agitated, hurt, angry or depressed. It wasn't serving me to watch the same movie on a daily basis. It was distracting me from all my areas of personal growth, and therefore working against my desired outcome. It kept me stuck.

My story had a place in my future, but only as a tool now. A tool to help others and to let them know that I too understand what devastation and relentless pain feel like. This tool would also allow me to give hope in the process of healing by sharing with others my personal navigation of it.

Once you gain your footing, you will have many people put into your path who need your insight, experience and hope. I saw this early on in my process. I remember about 5months post-divorce, I had recently committed to making hard cuts and on one particular day I found myself really struggling. Sunday evenings were the hardest for me as they were usually family days in my previous life. I now went to church alone and came home alone, especially since I was avoiding relationships and most social situations for a while in order to solidify my foundation. By the time evening rolled around, my sadness started sinking in and I couldn't help but cry. I wanted to medicate and make it go away, but again I chose not to. I chose the harder road in pure faith that it would eventually lead me to my light in the tunnel.

As I sat on my couch, by myself, my phone went off. It was an acquaintance asking me if I was home. I wanted to say "no" as I was in no mood to see anyone, but she insisted that she had a friend who was really struggling and who desperately wanted to talk to me. Knowing that, how could I say no? Not but 30minutes later, this woman who is a total stranger, was dropped off on my front porch. She had been in a very mentally abusive marriage, the worst I had heard of at that point and she was reaching out for help. Listening to this woman's story was devastating to me. In fact, my physical body started heating up internally almost, right in my upper stomach as I listened to the gut-wrenching details of what she had been through. I felt an almost PTSD response like I was reliving many ugly moments of my own life. She was incredibly thin, and I could tell that food restriction was how she masochistically had turned the pain inflicted by her abuser on herself instead. She was barely alive in a sense, metaphorically cutting her own wrists instead of standing up for herself. She was actually dying.

We talked for two solid hours. She needed someone to listen to her, who could understand as well. She needed someone to validate her sanity much as I had needed the night of my DUI. She needed to know that she wasn't crazy, and that she did not deserve what was being done to her. I gave her everything that I had and encouraged her to see her own value and strength. It was an intense evening, and I almost didn't respond to the message. An evening that had started with me marinating in my own pain and in my story, ended with purpose. My story had purpose and tonight, the frail woman sitting in front of me was it.

After she left, I did a lot of thinking and I accepted that I may never know exactly what path she chose to take thereafter, but either way, I was grateful for our moment. I was grateful that I had chosen not to grab a beer on the way home or avoid my phone message. Instead, I put my pain aside and gave of myself to someone sitting in the exact place that I had been not, but 7months prior.

Surviving the hard times takes a lot of courage and inner strength, both the things that may be lacking during times of excessive hurt, which is why we must keep believing in purpose even when we can't see it. That thing on my neck had rocked me to my core but ironically, had also improved life for me and my boys. You see, I had been living about an hour away from my family and friends for almost 10years. I had designed a life that consisted only of my husband and kids as well as my professional endeavors, so the distance wasn't too bad. We could easily visit family while maintaining our own world so to speak, as our own little family. After the divorce, I went on to live in the same area, near my boy's dads while focusing most of my life and time on my kids, and providing for them as well. I didn't really have a support system nearby, but it hadn't phased me since I had been so used to independently doing for myself. It wasn't until my medical scare that my perspective changed. Who would I call in case of an emergency? My ex and his girlfriend? No, thanks. What would I do out here if I were sick? The thoughts ran my already beaten down mind into the ground. And so, within 24hours I had made the decision to move closer to family and face whatever challenge that may befall me. I put a deposit on our new place, gave notice at our current residence and within 3 weeks, we were settled down in our own little community and much closer to friends and family. My oldest son enrolled in the best public school in the area, our place was much more spacious and it was really an improvement for us all. The first move was a sad move. This move was an upgrade; a new start of a happier life with our new family unit. If it weren't for my neck scare, I don't know that I would have ever even considered this as an option. I would have likely gone along, solo, doing what I do in the same environment that now housed a decade of painful memories. Instead, my scare had brought us closer to quality relationships and led us to a much needed, fresh start.

Finding purpose in pain instead of just focusing on the pain is a gift to ourselves. We desire to be happy and hopeful. Purpose gives us so much hope that even in our darkest hours, there will once again be light. I could have chosen to feel like a victim firstly to the divorce and now to this physical ailment but instead I was grateful. At the 11month mark post-divorce, my DUI was lifted from my record. My stellar record had earned me a withheld judgement, which meant that if I didn't violate the law during my probationary period, then it would virtually be removed. While it wasn't an eraser, it was as close as they come. I was absolutely ecstatic to hit that milestone and with all the remorse that accompanied my poor decision, I was even grateful for the DUI. At that point in my life, I was dying. My toxic relationship was killing me much like the woman whom I had later been connected to. It literally took that DUI for me to realize just how unhealthy I had become, and fight for the things that I truly valued; my children, my passion for helping people and my health. As that 1year mark came around, I embraced the whole experience; from painful beginning to painful end and to a new beginning. All was as it was always supposed to be. On the one-year anniversary of my DUI, I sat in extreme gratitude. I cried in gratitude and even with the unknowns of my future still lingering in my mind, I reminded myself to "trust the branch" as my son had so beautifully stated just months prior.

With gratitude comes even greater healing and as I was healed, I shed much of the resentment and anger which I had carried. While my ex had done a lot of damage, I saw him for the human that he was. His unresolved pain and wounds had created who he was, and what emanated from him. We only truly change when we are ready and open to it. Nobody can do it for us, and that marriage catapulted me into the most dramatic period of transformation in my entire life. If it weren't for that whole experience, I wouldn't be where I am today. I wouldn't have those experiences to source, or the words in this book. I may have never delved into my early life trauma and sought a new level of personal healing. The world was now happy in a way it had never been before and while there was no certainty in my future, I was very certain that it would always be okay. My Epic Year had not yet come to a close, but I could already feel the tides changing. I didn't know what exactly I would do with all of this, but I was open to my destiny.

I was so ready to live this new and improved life. While I had business and personal goals, I started intentionally feeling contentment right where I currently stood. I felt wealthy in life and started consciously giving in to enjoyment instead of constantly running. The constant run was like Indiana Jones when he's being chased by the boulder. You may have gold in hand, but if you can't stop to enjoy it, what's the point? We will never embrace or enjoy the blessings in our path if we never take a moment to absorb them. I still didn't have an answer for that thing on my neck, but had to decide that living with fear over it was definitely not serving me. My ability to release destructive thoughts and emotions is forever a work in progress, however, I feel that it was my greatest accomplishment throughout all of this. It took absolute torture for me to dig deep enough and decide to tackle a lifetime of excessive fear, and I was and still am truly grateful for everything that led me to do it. The pain that I went through was nothing compared to the chronic pain of constant anxiety.

It takes a tough shit to look at adversity with intention, but if you decide to, you will move much more swiftly through the hard paths of living this human life. I think that life is like a mountain with many peaks.

Just when we think that we have reached one, the fog clears and we see a much greater summit ahead of us. The climb will consist of peaks and valleys as well as flat terrain. It will also consist of impossible surfaces that aren't really impossible. You can and will find a way whether its up or through. Everyone is climbing their own mountain and at times, they may be right with you along your journey and other times, you will persevere alone. When you reach each summit, you will have two options; to carry on directly to the next or reach a hand to those below who just haven't quite found the right footing. This is your gift, but whatever you do, do not try to carry anyone else on your back. Your feet can lead masses, but your back can only carry few. Wherever you are today, know that your epic year of healing and purpose awaits you as soon as you decide that you are ready to embrace the entire process. Dually note that nobody else can carry you through it, or do it for you. Healing is done personally. It's your mountain to climb, and while you may not yet believe in your capabilities, if you stay in stride you will surely learn things about yourself and your true potential. Just keep moving on. Pressing forward is everything when it comes to personal growth.

You don't need the answers and you will likely have no idea what is ahead of you. You only truly need two things; the decision to start and absolute faith, and commitment to the process.

Chapter 9
A Warrior's Blueprint

"I named my oldest son "Phoenix" because it symbolized rebirth after death. Rising from the ashes. The power and growth that comes from destruction. The birth of a child is symbolic to a woman. We become warriors for a life more valuable than our own. Much like the transformation of childbirth, life continues to give us death and destruction so that we may grow stronger, wiser and rise from the ashes."
~Britt

If I have learned one thing all through these experiences; it is that moving forward can be excruciating at times and that motivation does not always accompany a desire for change. Many crave change because they desire to end their suffering even though they've experienced suffering for so long that they have no idea where to begin. While we all have a warrior or a "Red Night" within us, we still need actionable steps to effect that change we seek; a warrior's blueprint. History has it that great warriors didn't go to battle without a plan. They realized at some point that their girth and aggression wouldn't trump strategy, and that if they combine efforts, they would have a plan and the intensity to execute it. So, dear warrior, you need a blueprint.

I used this blueprint many times throughout my Epic Year and will still continue to use it in the future. In order to evolve, we need to heal and grow. Healing and growth require a multi-dimensional combination of mental, emotional, spiritual and physical focus. This has become the foundation for my Holistic Wellness Coaching Academy. The goal of a coach of any kind is to facilitate personal growth in any given area or multiple areas of their subjects. Holistic wellness is a circular connection between the mind, body, spirit, and emotions and each one feeds into the other. If we just focus on one area, we block the circuit so to speak. Make sense? So, if you are ready to create your epic year, be ready to go all in 100%.

There was no way a half-ass effort would have picked me up over and over throughout my darkest days. I became a fighter and I latched onto whatever I needed to. This included my students and anyone who looked up to me, my family who loved me and especially my boys who deserved a great life and a healthy mom capable of moving mountains for them. I would die for them but not like I had been. Not out of weakness. So rather than die, I fought for them and in turn fought for myself. I pulled from my pain like it was fuel for a jet engine and I used it as such. I used anger, resentment, sadness and especially fear. These emotions became the very blood that pumped through my veins and straight to my muscles when I needed to climb the mountain or pick myself up off the ground yet again.

It is important to state that not all of the hardships throughout this year were within my control. Our family saw 3 deaths in 5 months as well as health ailments and financial worries that ate deep into our subconscious mind. It was during these uncontrollable times that I really saw the purpose in my pain. Sometimes it was there to make me stronger for a future fight and other times, it was given to me so that I could help someone else. With that understanding, little that we experience in this life can break us because we truly believe that by experiencing it, we will become stronger than ever.

For clarification, this part of the book isn't a mechanical "how to" manual but an autobiographical account of the components involved in my own healing process and those who I have been blessed to mentor along the way. It's a major start on a journey to a life of healing and fulfillment. When your inner light is dim or your head isn't functioning properly, this strategy will make it easier to take on the big goal one basic piece at a time. The first essential step is deciding to begin. You deserve a happy, healthy, fulfilling life and I want nothing more than to help you create that life you envision. After all, my experiences as hard as they have been were for a purpose and this book shares major points of life. So now, let's find your purpose but first, we need a plan!

The Plan

Step 1. Omit toxicity

We all medicate with something; drugs, alcohol, sex, food, attention, work or relationships. It's a matter of acknowledging HOW we choose to medicate. Once you realize how you fill your voids or distractions from pain, you will have your first piece of the puzzle. Now that you know what your chosen medication is, you will get to decide whether or not to leave it behind. Get it out of your life as if it no longer serves the person you are becoming or the life which you are creating. It will handicap you from seeing life clearly as well as impede healing. I knew for a long while that alcohol needed to go as well as keeping people close who chose to reside frequently in that environment. Whenever I'm hurt or stressed, a drink was like a "vacation in a bottle," only I never came back refreshed. I was more worn out emotionally, mentally and physically. It only made everything harder and it seriously held me back. So, it had to get checked. The same thing was true for men as well. I was the girl who had been in one long unhealthy relationship after another and because there was no gap, there was no time to really find my best self. Who was I as an individual and what life had I ultimately wanted? I realized that if I did these two things first, then I could decide what kind of man complemented this life and deserved to be in mine and my son's environments. So, bye-bye boys! For the first time since I was 13, I chose to fly solo and honestly, it wasn't easy turning down dates and prospective "friends" at the onset, but I did it and after a while, the payoff became very clear. I started drawing in a new caliber of men and started truly feeling like I was deserving of a really great counterpart. To my own demise, I had drawn my personal value from a partner for so many years. Ironically, choosing not to have one was improving my self-worth more than anything else could have.

The path to a better life often starts as the seemingly "harder" road. Making the hard cuts that you have likely relied on for a long time won't feel easy initially. Just remember that it is so worth it, not in a ponies and rainbows kind of way but in a transformative, healthier, accomplished kind of way. A new level of enlightenment with far greater reward. The only way to do it is to just do it! Decide that the life you want is worth more to you than a momentary relief, which can be ultimately followed by destruction. Let go of who you have been and let go of anything that will keep you from becoming who you are meant to be.

Step 2: Live by your new routine

When you feel like shit, chances are that you won't necessarily feel motivated. Maybe you will! Whatever the case, count your blessings and roll with it, but routine will still be essential for you too. The next couple of steps will cover some important things to implement and it will be very important for you to start making these steps a consistent part of your daily life in order to get the greatest momentum. You may not feel like doing these things but do them anyways. Your future is worth it and a routine, almost brainless system will create the "fake it till you make it" consistency that will eventually undergo transition into a real change in mental clarity, wellbeing, motivation and eventually excitement. My life is all routine. Had I let go of that, I wouldn't have been able to take care of my kids or myself. Being a majority custody, single mom who chose not to take child support, I had a business to run in order to pay my bills and take care of my boys. If I went down, we went down and that was not an option. I set an alarm daily, moved my body every day, ate dinner with my boys at the table every night and so on. Your routine will become the foundation for your healing process.

Step 3: Eat, move, meditate, sleep

EAT

While it may seem trite to focus on food and movement as an intense aspect of personal healing, it is absolutely imperative. My first book "Buddha Belly, a mind, body, soul approach to health starting with your gut" was written after years of suffering from anxiety and depression. That suffering had led me to a new level of knowledge surrounding the importance of gut health and its connection to the brain, mood, and disease. Not only is a clean, nutritious diet important for overall health, the gut is also a direct link to your mood and it produces 90% of your serotonin and 50% of your dopamine, otherwise known as "happy hormones." The gut is where your emotional instincts lie, which makes it the foundation of your overall holistic healing process. While I can't go into intense detail about the gut in this book, you can find incredible guidance on gut centered nutrition online and of course in "Buddha Belly!"

Move

Whatever you do, don't stop moving. I mean that metaphorically as well as literally. Move your body every day and move forward in your healing consistently. Don't stop moving. If you wake up and have no motivation, put your shoes on and go for a walk. Whether rain or shine, just do it. A gym is just fine, but make sure to spend some time in nature. Nature has so many healing benefits; embrace them. Exercise increases oxygen and blood flow. It will help pick you up, especially when you feel like you are sinking or are stuck. Just move. If you can, set a physical goal of walking or running a 10k or take up yoga. Movement is a spiritual experience and it will assist in your healing process.

Meditate

Maybe meditation isn't your thing, that's okay. Making time for quiet mindfulness is very important. Make time daily for deep, oxygenating breathing, gratitude and affirmation. Recite powerful verses to yourself during this time such as, "I am growing stronger each day" or "I am absolutely enough!" Train your brain and give your body a chance to focus on its energy. As we discussed earlier, your subconscious mind is so powerful! Program it daily and you will start to see major changes.

Sleep

Adequate sleep cannot be understated. Lack of quality sleep robs the body of its "rest and digest" time, thereby predisposing it to not perform restorative functions and increasing cortisol, the stress hormone. Lack of sleep will only exacerbate anxiety, depression and mental well-being. This is a hard topic since the hormones responsible for inducing sleep are suppressed when stress hormones are high. It can sometimes feel like a catch twenty-two. The above-mentioned focuses aids in better quality sleep and you can even incorporate soothing teas such as chamomile before bed and avoid blue light from technology. Sleep should be a time of restoration and healing.

Step 4: Round up resources

Therapies such as counseling, alternative wellness treatments, and massage are an excellent aid. Navigating past trauma should be done with a professional who you can connect with and have trust in. You can and should interview any therapist whom you will potentially hire. Everyone practices differently and you will know when you have found your best fit. There are tons of alternative healing therapies, and while I can't recommend any specifically in this book, you can find many resources online or through our holistic academy at www.coachbrittsholistichub.com

Step 5: Build your community

You may need to make some hard relationship cuts. If your environment sets you up for failure, you may need to build a new circle of friends and community. If substance abuse is a problem for you and all of your peers are party people, this environment can easily hold you back from doing what is best for you. Don't be afraid to slowly create a new community. Join a healthy meetup group or start one! Get involved in a hobby and meet like-minded people or join a spiritual place of worship. Put yourself in healthy environments that feel comfortable to you and you will likely start building a more empowering circle.

Step 6: Run at your fear

Concur something. What is that very thing you've been too afraid to do? Maybe it's starting that dream business, jumping out of a plane or challenging yourself in some form of athletics. Decide on something and then run at it! Prove to yourself that YOU CAN achieve anything and that your previous limits no longer defines you as a person. If you do this in one area, it will bleed into every other area of your life. You will prove to yourself that you can achieve the impossible and that YOU hold all the POWER!

Step 7: Give of yourself

I can't stress enough how crucial this is. In fact, if you are really struggling and feeling lost, I recommend starting right here with step 7 and then going back to step 1. Giving of yourself through time, volunteering or anything that selflessly helps another is unbelievably healing. It doesn't just ease your personal wounds; it also increases your true value and changes how you see yourself. When we give selflessly to another, it comes from our soul and in doing so, we get a glimpse of just how beautiful that soul truly is. That is your true identity shining through.

Step 8: Passions and purpose

Find something that drives you. Is there any hobby or skillset that excites you? Find something that ignites the passion within you that isn't a person, substance or material posession. Passion will lead you closer and closer to purpose and purpose is what this life is all about.

Step 9: Growth. Seek it.

Keep growing. Even when you feel you have made headway in life, keep seeking out avenues for personal growth. Listen to motivational podcasts, read books, connect with individuals and be open to learning from others. I personally have an addiction for growth and I believe that this came from a sincere need to know that I wasn't stuck living with anything that didn't serve me. Whether it was a personal trait in myself or an experience that had previously haunted me, I gained power and assurance from knowing that I didn't have to live with that which hurt me. I could decide to grow from it and if need be, I'd outgrow it entirely. A lot of individuals fight their growth because it requires introspection and acknowledgement of personal realities that are often uncomfortable to face. Sometimes these realities are very painful but by humbly accepting our truths, we are given the opportunity for growth and ultimately gifted with rebirth.

Step 10: Allow yourself to live

Give yourself permission to live. Remember that you deserve a great life and that where you have been does not condemn your future. Your past experiences are merely your past story that paved the road that made you, and as you continue to grow, you will create purpose from the pain. Don't be afraid of being happy; you fought through the hard times, so now enjoy the clearing. The light at the end of the tunnel always awaits you. Give yourself permission to find it.

While any of these steps will benefit you, doing all of them consistently will transform you. You are created for greatness; a greatness that you will likely never fully understand. What you put into this world will cause a ripple effect long after you are gone. If we do not nurture ourselves and seek healing, that ripple can be quite toxic but if we do choose the path of growth, then we have an opportunity to leave the environment around us better than when we entered it. By healing ourselves, we raise healthier children who go on to become healthier adults. We all will reach that fork in the road with two options; to fight or die. Don't give up the fight or you may miss the incredible life awaiting you. It may just require a few steps further. Personal restoration is a process.

We did not choose many of the cards that were dealt to us and which contributed to our suffering, but we alone are the only ones that can change the trajectory.

The most transformative events for me took place in about a one year timeframe. I found this ironic since on a scientific level much of the body's cells are remade within one year! The Skin and stomach lining take days or weeks! Your body is constantly growing, improving, dying and regenerating. Not only should we assist this process for our physical well-being, we should also see this as a sign of transformation. We can completely grow into a new version of ourselves if we decide to take on the process. You're life and the purpose laid before you are worth growing for. This is your Epic Year! This is your time.

BRITTNEY OLIVER CHC

CHAPTER 10
FINDING THE LIGHT IN DARKNESS

I met Lori a few months into my Epic Year. I was contemplating selling my sauna and I posted it on a local for sale site. She inquired and came to look at it one fall evening. When she and her husband arrived, I couldn't help but notice how soft and sweet she was. She had informed me earlier on the phone that she was in the market for an infrared sauna to aid in her natural cancer therapies. They needed it because Lori was diagnosed with stage 4 cancer back in 2015. At that point, she had entered conventional cancer treatment and was assigned a short window of life. I didn't know what to say to her but I gave her a copy of my book "Buddha Belly" which I had sitting in an open box next to the sauna. I asked her to keep in touch as I would love to hear more about her story and natural approaches. She and her husband left that evening without purchasing the sauna and we did not remain in touch. About 3 months later, my boys and I moved about 25 minutes away, over the border from Idaho back into Washington State. I had taken the sauna off the site as I now had space for it in the new place. One day, I changed my mind and reposted it with no immediate intention of selling it. Another three months after the move, I received a call. It was a call from a woman who was looking for a sauna and had been battling stage 4 cancer. I asked if it was Lori and she sweetly informed me that it was her; the same woman whom I had met about six months prior. While I was happy to show her the equipment, I looked forward to connecting with her. Mind you, this was a smack in the middle of my neck scare. As I fought my own fears of terminal illness and death, I believed that my reconnection with Lori was no coincidence.

This time, I sold the sauna to the couple and requested that I interview Lori for my book. I made sure I conveyed the message to her with no pressures attached. I did not mention my personal situation, but I did tell her that her story would be a blessing for this book if she didn't mind sharing it. Lori agreed without hesitation to stop by again for tea and a chat. One beautiful sunny Saturday, we sat by the pool and we talked about life, death, treatments, mindset and more. Her perspective was incredibly insightful as she shared the rollercoaster that had been her life for the last four years. One thing I noticed was that she had no victim mentality. She saw her life as an experience which had been given to her. I asked her how she rides the ups and downs of bad days and good days and she answered my question more with body language than words. Her gesture communicated almost a statement. That statement was something like "Shoot, I don't know! You just do it and then do it again!" While Lori had her bad days, why was she sitting in front of me four years after a diagnosis that would have taken most individuals by this point? Was it her addition of holistic therapies and self-educated initiative over her care? That likely played a role in her survival but I believed her extended timeframe to be a product of something else. A combination of fight and acceptance. She fought against the typical medical approach and fought to keep her own mindset in a place of limitless possibilities instead of accepting someone else's timestamp. This was her life and hers to decide what to do with it. She also found acceptance in her story. She directed her fight at what she could control instead of fighting the things she couldn't. Her acceptance of the experience assigned to her gave her more power in the place of resentment and fear.

One major breakthrough moment that happened to me while talking to her was when she said "Ya know, sometimes I believe that the other side of life would be better than this one and because of that, I am no longer afraid." At this point, I asked her about the fear of death, something that had paralyzed me my entire life. She said that she had done a lot of work to get to a place where she no longer feared it, much like many cancer sufferers do and I absorbed that. Fear. Fear is what holds us back from our greatest dreams and it's also what keeps us in a state of misery. It's less about "the thing" and more about the <u>fear</u> of "the thing."

I asked her what her spiritual beliefs were and also made it known to her that I was no preacher nor would I ever judge, but coming from my own experiences, I had a curiosity for where others went when they were faced with a fearful or tragic experience. She had an interesting answer and preceded to tell me about her initial search for faith amongst religion in her early 20's, which led to a very unhealthy religious environment that eventually turned her away from spirituality for quite a while. Since the cancer, she had found a new depth of faith; one that still goes unspecified according to denominational religion. But she now knows and feels that no matter the day, or the point in time, that she is never alone. There is another energy in her presence at all times and when her day comes to advance from this material world, that presence will certainly accompany her to her destination.

I felt so blessed to be sitting in Lori's presence; to have had her brought into my life. She brought a story but more than that, she contributed a perspective; one of peace in the midst of fear. She embodied the reality of life. The entire process of life, growth and a wise understanding of it all. She too had found that fear was the greatest obstacle of this world and selfless giving was the greatest gift.

"One day closer"

Emily and I met back in 2016, just 3 months after the accident. We connected because of her charity Love 11 through my holistic network at the time. I reached out to Emily, offering to support some of her charity efforts with the profits from the network and we scheduled a day to meet. At the time, my husband and I met her and her husband for coffee at a local shop, and I approached the day with sensitivity knowing that these parents had lost a child. I did not know how but I had assumed that it was cancer or illness that had taken the life of their son and I honestly hadn't been prepared for their story nor the recent timeline. When we sat down and started chatting, they informed me that their son, Micah, had been taken in a tragic accident 3 months prior to our meeting. As a mother and a human, my heart broke for these people. They had faced what most of us would consider to be our biggest fear in life, yet here they were. Toward the end of this book, I knew that Emily's story of loss would be an impactful addition to this book and so I reached out to her without expectation and asked her if I could interview her so I can use her experience and Micah's life as a guiding light for others. She agreed.

Emily and Josh had gotten pregnant with Micah unexpectedly when they were still young. Emily at only 18 years old was determined to make a healthy life for this little baby and so she focused entirely on him, growing up quickly in many ways. The couple later married and lived as a family along with Josh's son from a previous relationship. The two tried for eight years after Micah to give him another sibling but to no avail. After many doctor's appointments and testing, they were left with no explanation as to why they could not conceive. While still trying for another child, they were grateful for the ones they were given unexpectedly.

Micah was a happy and selfless little boy who constantly befriended the underdog. He had a big heart and a stubborn mind as all the most impactful people do and it was clear that he would be a leader in things bigger than himself. Emily recalls memorial weekend of 2016 as a chaotic blur that would forever change her life. After reluctantly agreeing to allow Micah to attend a family camping trip that she and Josh would not attend, she sent him off to spend a few days, so he could enjoy nature with some of her relatives. Minutes after Micah had arrived at the campsite, he wandered over toward his grandpa and a couple of others who unbeknownst to him were falling a tree for firewood. There was a storm coming and a few of the family members were cutting down wood to get them through what was looking to be a rough night.

Nobody saw Micah coming until the large tree was in mid-fall. Emily's dad explained that moment as if it were in slow motion. He remembers locking eyes with Micah and saying "run!" But it was too late. The tree fell on the small-framed eight-year-old. The family acted immediately by calling 911 who eventually responded.

Emily and Josh got that dreaded call and ran frantically for multiple hours tracking down Micah's whereabouts and trying to reach the destination where he would be taken. After almost 5 hours, Emily recalled the phone conversation that changed everything. The doctor working on Micah got on the phone and said, "We are preforming CPR and are about to call it off." Emily and Josh begged the doctor to keep going and at that point, the doctor bluntly stated, "Your son will be dead when you arrive here. You need to prepare for that."

As Emily retold her story of that moment, my gut wrenched with empathy, love, and fear. As a mother, how would I handle this? How did she? She said that she and her husband sat with Micah's body for quite a while after arriving, not wanting to leave him. Not wanting to leave their baby. While his spirit no longer occupied his physical body, they describe feeling his presence in the room with them, being with them as they processed life without Micah in this material world.

The couple described the first six months after Micah's passing as living in a haze; a world of their own. It sounded similar to my "dome" that I lived in for some time; where you can see everyone looking in on you, but no one can reach you. It is just about you and God sitting together and going through the process. The difference is, these parents were going to emerge into a new world as new people without their child. As many of us can empathize with, they both spent many days contemplating exiting this world in order to be with their son and to escape the pain. Emily's favorite saying given to her by her husband is "One day closer" as this reminds her to keep going. It reminds her that this life is temporary and that she will be reunited with her child someday.

In the meantime, this couple started a charity foundation in Micah's name with his life insurance money. Their charity Love 11 gives sports scholarships to low-income youth allowing them to participate in sports activities and community. The Love 11 foundation represents a loving little boy who reached out to those who needed a friend and a positive mentor. Emily rides the waves daily from struggle to resilience, yet she keeps going, finding fulfillment in children and travel and knowing that each day is an opportunity to expand Micah's legacy. Each day she is "one day closer." For every day that Emily and Josh live this unimaginable journey which they were given, they extend the ripple effect of Micha's life both with the charity reach and through using their story to connect to others in pain.

After interviewing both Lori and Emily, I was overwhelmed with emotion and gratitude. The impact of their stories knocked me out of reality a bit. I paused, thought and felt absolutely grateful to be the one allowed to write their stories in this book. One of these women was grateful for each day forward and the other was counting down "one day closer" to the next life. Both have a perspective that we all need to hear. They see life as temporary and meaningful. They help remind us that substance and relationship are what makes life worth living and that all of the material focuses only distract us from the substance. They also represent most people's greatest fears; terminal illness and loss of a child. And yet, here they are. So, what resides on the other side of fear when you have lived your greatest fear? Throughout my hardest days, I strongly identified my need to process through fear. What I found to be my only real tool for fear was faith. Lori has faith that she is never alone and will rest in the light of peace and Emily has the same belief. While the timelines for each of us are unknown, I have come to believe that faith in an ultimate peace can trump any level of fear or pain. I sensed that these two women would agree.

How could three different people from three completely different challenging life circumstances come to the same conclusion? Both Emily and Lori discussed navigating fear and hopelessness and they both shared their innate love of giving to others. I agreed completely. My year had walked me through a long dark tunnel to navigate my own relationship with fear and somehow I stood there in the light. I did not stand in the light because of the outcome but rather the perspective, just as these women did. With Lori battling terminal cancer and Emily having lost her son, both saw the light awaiting them and they keep walking even when it becomes too dark to see. Their stories and their journeys are light in its finest sense.

Watching these women live in purpose, fight and faith is breathtaking and humbling. How many of us go through our daily lives focused on the material world around us and being disconnected with our deepest self? Our greater purpose. Imagine all that we can do in this world if we only tap into those who have been thrown into a deeper level of existence. My intention in sharing these two stories is to reach out to those in darkness.

Not every story has a "happy ending." Emily will never say that she "wouldn't change a thing" as I would say about my situation, and I am sure that Lori wouldn't say so either, but both women have taken the cards dealt to them and have pressed on. They contribute love to the world around them and they connect with those in pain too. They used their own pain and experiences to fuel their purpose in this life, and for that, they are a representation to every human. They represent warriors.

Chapter 11
Fire Walk

Pain has a brilliant design in that it builds spiritual muscle. The saying, "what doesn't kill you makes you stronger" is a beautiful truth. Often, weakness is a result of fear and if you have been through hell, what then is left to fear? Without fear, we become unstoppable and aggressive in the pursuit of our dreams and purpose. Sometimes the very fight in us that is necessary for our survival becomes the aggression needed to catapult us to a greater consciousness. This life is a journey and the intense periods are your "fire walk." A fire walk isn't easy, but it is transformative. People don't engage in fire walking because it's an easy thing to do. They do it because it tests their humanness by testing their limits; the limits of their physical body and their mind along with their beliefs. By pushing past our beliefs of what we "can" or "can't" do, we find our personal power. We are so much stronger than we really know and while we are made of flesh, our energy is something else entirely.

After leaving my marriage, I was broken. The DUI made it clear that I had been broken for a long time. Like we discussed earlier, healing ebbs and flows, and we sometimes gain the momentum only to backpedal shortly thereafter. What matters is that we keep going and believing in faith. During this year of my life, I backpedaled more than once. I felt like I spent a year picking my ass up off the ground, bloody, broken and weak but I still kept getting back up and I never lost faith that "this too shall pass." I knew that this year was a fire walk and a fire walk meant transformation. About 6 months in, I remember thinking "I wonder who I will be after all of this" and I wondered about what was yet to come. It was as though there were an innate guidance foretelling my future in a vague and suggestive fashion, but with the reassurance that a great transformation was coming with a great purpose. And so, I just kept getting back up. This was my year to transform more than ever in a very short time. That perspective is what I credit to my success and survival. Without my unrelenting faith in the purpose behind all of this, I may have lost it all; my sanity, my health, my stability and the life which I had dreamed for my children. The one thing that became the fork in the road and the difference between two polar opposite outcomes was my rock-solid faith. I believe that God molds us for purpose and we only continue to suffer if we resist the transformation, which entails resisting growth. Growth is uncomfortable but stagnation is deadly. I refused to let my life become stagnant.

I sort to envision a fork in the road similar to being dropped in an open field of battle during medieval times and handed a spear. You are on one side and you have two options; fight or wait to die. Nobody cares whether or not you have trained your whole life for this moment or whether you were just placed here unexpectedly. Your options are still the same. So, what would you do? You can't run since the opposing side is too close and the field is too open. You either wait to be impaled or you grasp your spear, let out a powerful charge voice and run straight at the very thing that wants to kill you. Life is very similar. Sometimes, whether we are ready or not, it comes and presents two choices; fight or die.

I have a funny story about this actually. When I was about 5 years old, the neighbor two doors down had a terrifying and vicious dog. Looking back now, it was basically a crazy puppy but as children, my sister and I were scared shitless of this creature. We wouldn't even go outside if we thought it was out there and we would often play on our porch for fear that it would be let loose at some point. Sure enough, that day came. This monster dog got out of its yard and charged straight towards us. At the time, I was 5 and luckily for me, I was on the porch by the door which allowed for a swift getaway but my 3 years old little sister was not. She stood in the middle of the yard oblivious of her oncoming fate. I had two options; save myself and sacrifice my sister or attack the dog. Before I even realized what was happening, I became that random warrior in a field charging the enemy with a spear in hand. Instead of a spear, I picked up a stick. A sound came out of me at that moment that I had never before heard and I charged the beast. Much to my surprise, the dog stopped out of a full barking sprint, turned around and ran back scared to its yard. My intensity had caused the enemy to retreat and my sister was now safe. The best part was that my dad had been watching this whole thing unfold through the window. I was mortified as I had hoped that nobody had seen what had just happened. I was not in my right mind! When I walked into the house, I was greeted with a very serious slow clap and presented with a verbal medal of honor for my bravery. Much to my embarrassment, he still tells that story to this day.

That story is eerily similar to my reoccurring dream about running at the demon in the room. For a child who had experienced such fear early on in life, who grew into an adult still struggling, how strange were these dreams and moments of bravery? Were they really bravery or were they my subconscious, God-given purpose forcing me through my barriers? Whether they were out of desperation or being extremely fed up, they were buried in me underneath massive fear and disguised as weakness.

As the year wrapped up, a clarity had come to me. Fear was my mountain. Fear had incapacitated me since childhood and into my adult life. It held me back from life and tortured me for as long as I could remember, especially my fear of the unknown. The week prior to my one-year DUI anniversary, I had to go in for a CT scan on my neck for what the doctor was confident was no big deal. While I had not counted on this outcome completely, I had been looking forward to closing that chapter. I found it a bit overwhelming that the door would close on my 5-month long health scare simultaneously alongside my DUI being lifted. I was ready to run at life completely now. By this time, I had become happier and healthier than ever before and I did it while flying completely solo after massive destruction.

I awaited my phone call and when it came, it did not bring the closure that I had hoped; the closure that I had so humanly planned would be my fate. The scan showed something different than what was originally expected and now they ordered an MRI. While the practitioner seemed somewhat comforting that the outcome would yield a less serious result, I initially could not find comfort in his words. I worked in holistic health and understood enough about conventional medicine to know their protocol. I knew that the doctor was hoping to put the worst-case scenario to rest with the MRI but if not, the next step of "drawing fluid" meant testing for cancer. Instead of having the gates fly open to my new, powerful life, I felt the crushing blow of heavy fear once again in me. Three months prior, I had been on my knees begging God to lift the weight of this experience and give me a favorable result so I could really live the life I had worked so hard to create and heal for. My DUI door was closing, and I was developing an amazing community of new friends, people, and relationships and I was really truly happy. Now, it was all set aside and I was terrified. I knew that I would face weeks of uncertainty as the MRI would need to go through insurance approval, and instead of celebrating my one-year transformation, I was waiting for my next step.

The unknown is often one of the scariest places to reside. We fight uncertainty daily by trying to control or prepare for events in our mind through playing out scenarios, planning for the "worst case" or often with deflection or denial. How often though do we sit in uncertainty and just give in. I got to a point where I realized that giving in was all that I could do at that point. In a weird way, I felt like I was back at square one that night of my DUI and reliving my trust fall with God; letting go and trusting only His arms in place of mine since I had nothing left. I knew that while I prayed for a favorable outcome, either way, I would write the end of my story. If God closed the door on this life or if the path was destined to get darker before the light, that I would trust the purpose which He had laid for me. Either way, I believed that He didn't give me any of this as punishment but rather as a story that needed to be given to others.

None of us know what our story is meant to become. The power is in how we decide to use it. I knew that I would use my story and that I wasn't meant to waste it on self-pity or melancholy. If we can't change our circumstances, we owe it to ourselves to use the scraps of broken glass to create a masterpiece that nobody has set eyes on before. It is original and new to this world because it is molded from you and your story; your experience and your purpose. The world needs to see your masterpiece through your eyes to gain a new perspective of their own life and circumstances.

As I reflected on my year and essentially my entire 34 years of life, I remembered vividly the feeling of reassurance. I remember feeling like something greater than myself was preparing me for an intense transformation and that this would only be possible with great pain as well as incredible blessings, faith, and gratitude. It's like I had known innately from the beginning that the road would get much more treacherous before it would reach the clearing and that in uncertainty, I needed to give in to the process. While I still sat in uncertainty, I couldn't help but feel that I was nearing the end of the road. The bright light I had envisioned begin warming the path for me no matter what the future held. In fact, during my actual MRI, which if you've never experienced those, its a large, loud tube you get tightly inserted into and are not allowed to move for at least 30 minutes, I basked in gratitude. I was oddly overwhelmed with emotion that day as I reflected on so many events of the earlier year and up to that moment. I couldn't have possibly foreseen the immense, chronological events which would lay out in front of me, and poignantly, I might add that it painted almost irrefutable evidence in the existence of a mighty power far greater than myself. One who designs a path of purpose for us through twists and turns, darkness and light and uses us to write a story.

My story through my life, especially during what I now call my Epic Year was a big one. This year was a big one and honestly, I had asked for every bit of it. I had chosen many of the things that led me to where I was in one way or another and now in this year, I had chosen to write my own ending. I had asked for all of the events which had played out before me and it all started that warm night in May when I was drunk in my car, I parked in a dark parking lot and begged God to save me. We have so many notions as to what answered prayers should look like. We assume that if they don't go exactly as we have envisioned in our minds, that we were ignored. The truth I believe is that God knows exactly how to lead us to our own healing and peace, and the trick it to trust Him with all our hearts. Again, I didn't write this book to push a religion and while it may be off-putting to some believers, I am not offended by what higher power another feels connected to. With that said, I must be authentic and true to myself and my story. This journey was mine to walk alone, but I was never truly alone nor could I have gotten to the place that I am now if I had been truly alone.

From the moment that I broke down, everything put forth in front of me was set to mold a new version of myself that the world needed. That woman broke so that a new woman could emerge and carry out a greater purpose; from the DUI to the humbling legal and labor experiences that followed. From the divorce and the pain of single motherhood to the beauty of a powerful independent life and healthier, happier children. I could see now, how all of my prayers had been answered. In the MRI machine, I tried so hard not to cry. I was so overwhelmed with emotion as I watched the movie which had been my life virtually played out before me.

At the beginning of this year, I was a broken woman in an unhealthy marriage, whose kids were paying the price. My business and my passion were more of a hobby and my health was not where it should be for my profession. I had very little to excite me out of bed each morning and I spent more time in my partner's head and thoughts than my own. I wasn't living my life; I was surviving it. While I laid in that MRI machine, I had so much to be grateful for in such a short period of time. My kids were happy in our new life and I had truly found contentment there too. I had grown my business and connected to people and relationships in my professional arena I would never have believed was possible! I reached goals and even had gotten into the best shape of my life. Just a few weeks prior, I met a great man who had become an ever-growing companion. It was a meeting I did not plan, did not expect and it superseded my previous beliefs of men. I reconnected with family I hadn't seen since I was 7 and I also met new relatives for the first time. While I knew better than to put my strength or happiness into the hands of others, I had started to understand that allowing others in was an important part of my healing process. I was at a very vulnerable place, uncertain of my future and my health and here, I had been given these special people who insisted on standing by my side in the darkness. I had always considered myself "like a cat." If I was wounded, dying, suffering or in pain, I would hide away alone in fear and tend to my own wounds. I didn't know how to let anyone else in there, yet here I was with a circle around me and I was so grateful.

I was anxiously awaiting my MRI results and had called almost daily for 2 weeks. I tried to put it all out of my mind but that was a challenge even at my best mental state. One particular day when I was really struggling, I snuck away from where my boys were sitting, so I could to talk to God. I was so tired mentally and emotionally and I was so ready to move forward with whatever may come my way, I just needed the go. In that moment, a crazy thing happened. It was as if a huge lightbulb turned on in me. All of a sudden, I was gifted an answer about not only this current situation or even this year but an answer to my whole life. There was a BIG piece of me that needed healing and that part of me only knew how to cope alone. We weren't meant to be completely alone. We are gifted family, friends or even community for a reason and with the energy and support of others alongside with our faith, we don't have to live in fear. At that moment, I realized that I wasn't alone. I was blessed with love in the form of close friends, family and even new relationships, and these people closed in tightly around me as I walked the last leg of my darkest hour. While I was strong and independent, my inability to let others in was a major weakness that manifested because of the deep-rooted, early wounds that I wasn't meant to carry anymore as they would no longer serve me.

I cried. I felt that I was crying out the weight of 34 years, just like the end of a movie when the music escalates to emphasize the finale. Then I heard it, literally right at that moment, the phone rang. The phone call I had been waiting for, for two weeks. It was my doctor informing me that while my neck thing was an odd and still challenging condition, it was not life-threatening and that they would schedule another MRI in 6 months. And with that, my Epic Year would come to a close. Not only could I not have predicted the last year's events, I couldn't have possibly anticipated that all doors would close within weeks almost one year to the day from my initial pleading with God.

My story is not unlike your story. I looked back on the previous year's events; all of the trials and the successes an I honestly felt blessed. Overall, I'd do it all again. I would relive the hardest, scariest and most painful year of my life in order to stand where I do now, as who I am today. I chose to embrace the transformation of life in all of its seemingly callus, beautiful and ironically powerful ways. While I was certain that my year was over, I embraced the uncertainty of the future awaiting me. I left a lot of detail out of this book such as some of the major events or situations that happened to me. My reasoning for this was not only out of protection of relationships and people, but for the fact that we all have a story of pain in some shape or form. Not one of us will get out of this life without suffering. The gift of life is in the power which we hold over ourselves. We can decide to change our story, to use our suffering to impact lives and to grow from anything. We can create the life we desire instead of seeing ourselves as victims to that which is out of our control because once we start that journey, we will always find light in purpose; a place where we are needed, where we are strong and where we are so powerful. If you feel the heat of life, see it as your fire walk and embrace it. This may be the beginning of your Epic Year and if so, embrace it with swords drawn and an open heart to become more than you could imagine.

While writing this book, I knew what I wanted for the cover. I called up a friend of mine who is a genius with conveying emotion through photo and video and described my vision. I wanted to be in the pic, standing on top of a mountain. I wanted this literal representation of a very metaphorical experience in my life and he absolutely delivered. When I had first brought up the idea, he suggested holding the shoot somewhere impactful such as Glacier National Park. I laughed at the irony of all of this since he had no idea of my previous years plan to hike the park and gain closure of my divorce, only to have it burn down in front of me. So, we planned it. The shoot was awesome and we couldn't have picked a better time and place. I stood out there, looking around, absorbing everything from the previous months and feeling overwhelmed with emotion and gratitude for it all. I had scaled the biggest mountain of my life and the fight was absolutely worth the person I had become today, the relationships and the purpose which I had found.

Our second night in the park, I received a text message from my mom, informing me that a letter had arrived for me at their house. I asked her to open it and send me a pic. When the message came through, I couldn't believe it. It was the letter that I had written to myself at my Tony Robbins event, 5 months prior.

A very broken woman had written that letter and I remember at the time, I felt like the words were so far from reality. I was writing to my future self that day in hopes to stand where I was in this very moment. As I read the letter out loud to my friends, I cried out knowing that God had brought me exactly to where He wanted me. The darkness was lifted, I had made it and here and I was humbled by the light.

The letter read:

"Dear Britt,

Your passion, your "why" is bigger than any roadblock or failure and for that reason, you will succeed. You have closed a big chapter in your life. Be proud of the warrior which you have become and never forget that you hold more power than you sometimes believe. This power will transcend fear. Keep your heart right and God will continue to light the way. Use your gifts and be willing to accept all that will come with your future. Nobody can touch that which you have become. Never forget to live, love and be grateful.

You are winning at life!

Love,

"Red"

Thank you to all who contributed to the creation of this book! All proceeds from the sale of #MYEPICYEAR will go directly to Coach Britt's philanthropic focuses and documented on her personal site. Interested in becoming a certified holistic coach, following Coach Britt or giving to Love11 in support of Micah's legacy? See the resource list below.

Resource List

- Follow Coach Britt for event info and to follow her philanthropic adventures: www.coachbrittsholistichub.com
- Instagram: buddhabellycoach
- Facebook: Coach Britt's Holistic Hub
- Info on becoming a certified coach: Holistic Wellness Coaching Academy www.hwcacoach.com
- Instagram: coaching.school
- Facebook: Holistic Wellness Coaching Academy
- Love11 charity foundation: www.love11.org
- Photo credit: TG Squared Productions www.tgsquaredproductions.com

Made in the USA
Columbia, SC
09 August 2020